The History of Science
Volume 6
The Eighteenth Century

Dr. Peter Whitfield

First published in the United States in 2003 by Grolier Educational, a division of Scholastic Library Publishing, Sherman Turnpike, Danbury, CT 06816

For Compendium Publishing
Contributors: Sandra Forty
Editor: Felicity Glen
Picture research: Peter Whitfield and Simon Forty
Design: Frank Ainscough/Compendium Design
Artwork: Mark Franklin/Flatt Art

Reproduced by: P.T. Repro Multi Warna, Indonesia.

Printed in China by: Printworks Int. Ltd

Library of Congress Cataloging-in-Publication Data
Whitfield, Peter, Dr.
History of science / Peter Whitfield
        p. cm.
    Includes index.
    Contents: v. 1 Science in ancient civilizations – v. 2 Islamic and western medieval science – v. 3 Traditions of science outside Europe – v. 4 The European Renaissance – v. 5 The Scientific Revolution – v. 6 The eighteenth century – v. 7 Physical Science in the nineteenth century – v. 8 Biology and Geology in the nineteenth century – v. 9 Atoms and galaxies : modern physical science – v. 10 Twentieth-century life sciences.
    ISBN 0-7172-5729-0 (act : alk. paper) – ISBN 0-7172-5703-7 (v. 1 : alk paper) – ISBN 0-7172-5704-5 (v. 2 : alk paper) – ISBN 0-7172-5705-3 (v. 3 : alk paper) – ISBN 0-7172-5706-1 (v. 4 : alk paper) – ISBN 0-7172-5707-X (v. 5 : alk paper) – ISBN 0-7172-5708-8 (v. 6 : alk paper) – ISBN 0-7172-5709-6 (v. 7 : alk paper) – ISBN 0-7172-5710-X (v. 8 : alk paper) – ISBN 0-7172-5711-8 (v. 9 : alk paper) – ISBN 0-7172-5712-6 (v. 10 : alk paper)
    1. Science–History–Juvenile literature. [1. Science–History.] I. Grolier Educational (Firm) II. Title.
Q125 .W586 2003
509—dc21
                                                        2002029844

Acknowledgments
The publishers would like to thank the following for their help with the illustrations: Venita Paul and Sarah Sykes at the Science & Society Picture Library, Science Museum Exhibition Road, London SW7 2DD; and Dawn Hathaway and Hillary Smith at the Natural History Museum (**NHM**— nhmp@nhm.ac.uk 0044 (0)2079425401).

Picture credits
All maps and artwork are by Mark Franklin/Flatt Art.
All photographs were supplied by the Science & Society Picture Library except those on the following pages (T=Top; C=Center; B=Below): Author's collection p.8, p.12, p.17, p.19 (B), p.30, p.31, p.46, p.52, p.72; NHM p.3 (B), p.16, p.19 (T), p.20, p.21, p.22, p.23, p.44; p.57 (NASA/Science & Society Picture Library), p.69 (National Railway Museum/Science & Society Picture Library).

Note
Underlined words in the text of this volume and other volumes in the set are explained in the Glossary on page 70.

Contents

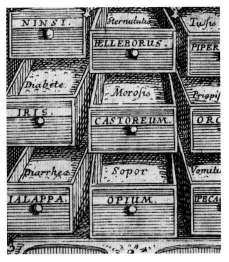

Introduction:
Science in the Eighteenth Century

The great scientists of the seventeenth century, from Tycho to Newton, had initiated not merely a revolution in science but a revolution in thought. This revolution created new standards of knowledge and of truth because it insisted on new ways of looking at the universe—by observing, measuring, and analyzing the physical forces that governed it.

Before the year 1600 the Western intellectual tradition had consisted of the study of classical literature, the arguments of classical philosophy, and Christian theology. A century later, however, after the year 1700, no educated person could remain ignorant of ideas such as Copernican astronomy, the theory of gravity, concepts such as chemical reaction, magnetic force, the vacuum, and the hidden world of microscopic life. Public lectures and experiments even became a fashionable form of entertainment.

There was, as yet, no reform of the educational system to spread and develop these ideas. The universities of Europe remained wedded to an almost medieval syllabus of classical literature, and the only science taught was mathematics. It was the scientific

Alessandro Volta demonstrates his "voltaic pile"—an early electrical battery—to Napoleon.

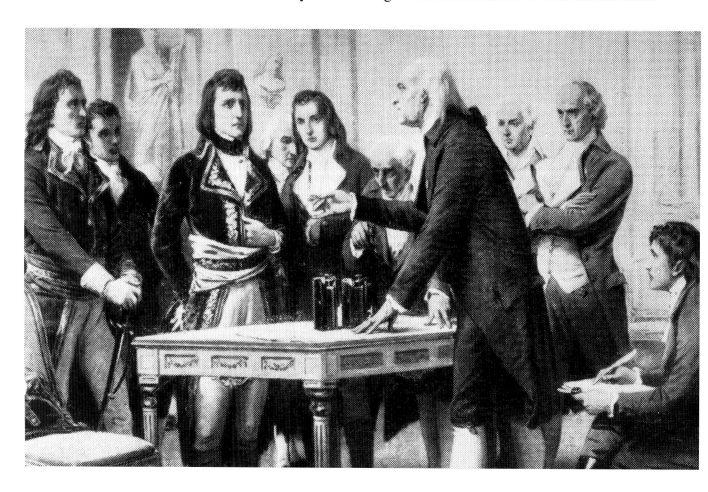

academies established in the major cities of Europe and, later, in North America that gave science a visible presence in society and provided scientists with a forum in which to exchange data and ideas.

Nor had science yet made any great impact on the way society functioned, for the machine age still lay in the future. A technological revolution was in the making in eighteenth-century Britain in the form of steam power, but it had a craft basis, and its pioneers were not pure scientists. The steam engines were built in mining regions, and their significance for the future was not yet grasped; a landmark study of economics such as Adam Smith's *The Wealth of Nations,* published in 1776, made no mention at all of steam power.

A FRESH APPROACH

How then did the scientists of the eighteenth century build on the achievements of the scientific revolution? Their work advanced in four main directions. First, there was an intense development in mathematics, both pure and applied. A great deal of this work was aimed at refining Newton's dynamics, analyzing the movements of the heavenly bodies in ever-greater detail in order to establish that gravity really was the universal force that held all things together. Second, experiments with physical forces continued, resulting in the discovery of electricity. Third, there was a huge accumulation of

Above Left: Orreries—working models of the solar system—seem to embody the eighteenth-century faith in the orderly mechanical universe. The one at left has armillary bands, was made between 1740 and 1747, and shows the planets out to Saturn. It was probably made by Thomas Wright (see below, page 57).

Above: Tiny celestial globe (with a three-inch diameter) on a brass mechanical stand for demonstrating the motion of the moon. It dates to the mid-eighteenth century.

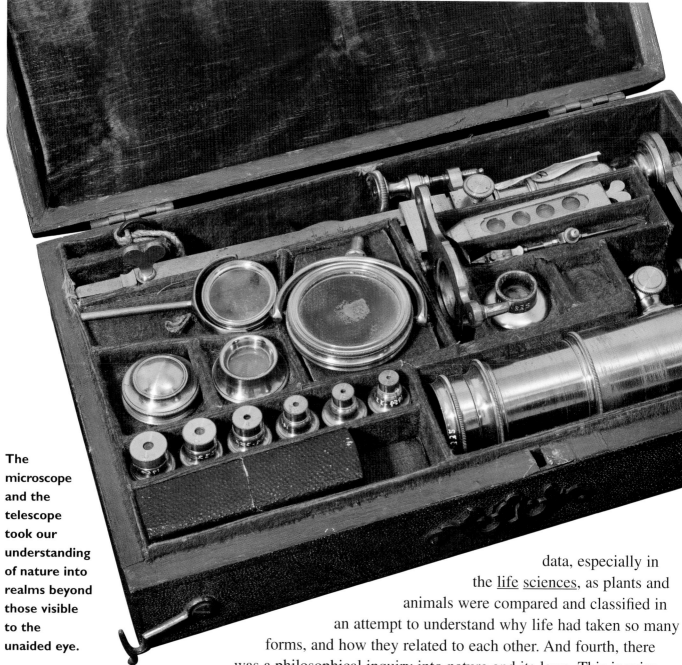

The microscope and the telescope took our understanding of nature into realms beyond those visible to the unaided eye.

Above: This is Joseph Priestley's microscope of 1767 (see Volume 7, page 24).

Opposite, Above: "Viewing the transit of Venus," an engraving of 1793 published by Robert Sayer & Co.

Opposite, Below: The first Gregorian reflecting telescope, made by John Hadley in 1728. This type of telescope uses two concave mirrors and was proposed by James Gregory of Aberdeen, Scotland, in his book *Optica Promota*, published in 1663.

data, especially in the life sciences, as plants and animals were compared and classified in an attempt to understand why life had taken so many forms, and how they related to each other. And fourth, there was a philosophical inquiry into nature and its laws. This inquiry attempted to spell out the implications of the new science, and what it meant for traditional ideas about God and humans. It reached different conclusions in England and in France. In England science was held to support a religious view of nature, while in France it produced a more critical spirit in which prescientific beliefs were dismissed as superstition; atheism emerged in France long before it did so in England.

The eighteenth century saw fewer landmark discoveries than the seventeenth century, but it was a period in which an awareness of science spread deeply among educated people. Newtonian physics, in particular, became a vital part of the accepted belief system of the age. We regard the eighteenth century as an age of reason, of enlightenment, and the rise and spread of scientific thought was a major component in that cult of reason.

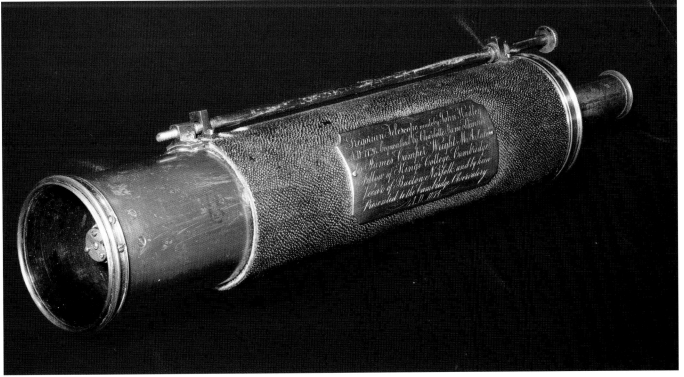

Mr WHISTON'S SCHEME of the SOLAR SYSTEM EPITOMIS'D. To wch is annex'd.
A Translation of part of ye General *Scholium* at ye end of ye second Edition of St *Isaac Newton's* Principia. *Concerning God.*

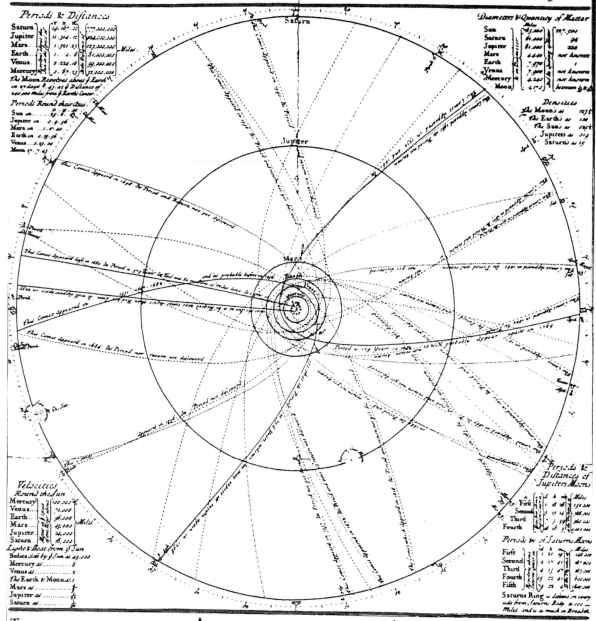

THIS most Elegant System of the Planets and Comets could not be produced but by and under the Contrivance and Dominion of an Intelligent and Powerful Being. And if the Fixed Stars are the Centers of such other Systems, all these being Framed by the like Council will be Subject to the Dominion of *One*, especially seeing the Light of the Fixed Stars is of the same Nature with that of the Sun, and the Light of all these Systems passes mutually from one to another. He governs all things, not as ye Soul of the World, but as the Lord of the Universe, and because of his Dominion he is wont to be called Lord God παντοκρατωρ (i. e. Universal Emperor) for God is a Relative word, and has Relation to Servants: And the Deity is the Empire of God, not over his own Body (as is the Opinion of those, who make him the Soul of the World) but over his Servants. The *Supreme God* is a Being Eternal, Infinite, Absolutely Perfect; but a Being however Perfect without Dominion, is not *Lord God*: For we say, *my God, your God, the God of Israel*, but we do not say, *my Eternal, your Eternal, the Eternal of Israel*; we do not say *my Infinite, your Infinite, the Infinite of Israel*; we do not say *my Perfect, your Perfect, the Perfect of Israel*. These Titles have no Relation to Servants. The word God frequently signifies Lord, but every Lord is not God. The Empire of a Spiritual being constitutes God true

Empire constitutes True God, Supreme the Supreme, Feigned the Feigned. And from his true Empire it follows that the true God is Living, Intelligent & Powerful, from his other Perfections that he is the Supreme or Supremely Perfect. He is *Eternal & Infinite, Omnipotent* and *Omnipresent* that is, he endures from Eternity to Eternity, and he is present from Infinity to Infinity, he Governs all Things, and Knows all Things which are or which can be known. He is not Eternity or Infinity, but he is Eternal and Infinite, he is not Duration or Space, but he Endures and is Present. He endures always and is present everywhere and by existing always and every where, he Constitutes Duration and Space, Eternity and Infinity. Whereas every Particle of Space is *always*, and every Individual Moment of Duration is *every where*, certainly the Framer and Lord of the Universe shall not be (*nunquam nusquam*) never nowhere. He is Omnipresent not Virtually only, but also Substantially, for Power without Substance cannot Subsist. In him are contain'd and moved all things (so the Antients thought †) but without mutual Passion God suffers nothing from the Motions of Bodies. Nor do they suffer any Resistance from the Omnipresence of God. It is confess'd that the Supreme God exists Necessarily, and by the

same Necessity he is *always* and *every where*. Whence also he is wholly Similar, all Eye, all Ear, all Brain, all Arm, all the Power of Perceiving Understanding and Acting. But after a manner not at all Corporeal, after a manner not like that of Men, after a manner wholly to us unknown. As a Blind Man has no notion of Colours, so neither have we any notion of the manner how the most Wise God perceives and understands all things. He is wholly destitute of all Body and of all Bodily shape, and therefore cannot be seen, heard, nor touched; nor ought to be Worshiped under the Representation of any thing Corporeal. We have Ideas of his Attributes, but we know not at all what is the Substance of any thing whatever. We see only the Figures and Colours of Bodies, we hear only Sounds, we touch only the outward Surfaces, we smell only Odours, and tast Tasts, but we know not by any sence or reflex Act the inward Substances, and much less have we any Notion of the Substance of God. We know him only by his Properties and Attributes, and by the most Wise and Excellent Structure of things, and by Final Causes, but we Adore and Worship him on account of his Dominion. For God without Dominion, Providence & Final Causes is nothing else but Fate and Nature

Engrav'd and Sold by John Senex at the Globe in Salisbury Court near Fleetstreet. Where may be had Dr Halleys Scheme of the Total Eclipse of the Sun which Shall be in 1724. Also his Zodiack containing all the Stars in the Way of the Moon and Planets, useful in observations and to find the Longitude at Sea.

The Vindication of Newtonianism

At the time of Newton's death in 1727 the prestige of his science was immense throughout most of Europe, but there were still doubters and opponents, especially in France, where Descartes's <u>mechanistic theories</u> were preferred. In the course of the eighteenth century a great deal of theoretical and practical work was devoted to the business of proving or disproving certain features of Newton's theory of gravity.

OPPOSING VIEWS

Perhaps the most impressive practical demonstration concerned the shape of the Earth. Newton had predicted that the Earth could not be a perfect sphere, but must be slightly flattened at the poles and slightly bulging at the equator. This was because the Earth's rotational speed was highest around the equator, and <u>centrifugal force</u> strongest. Descartes and his followers argued, on the contrary, that the Earth must be elongated toward the poles because it was squeezed by the <u>vortices</u> that they believed swept the Earth around in its orbit.

It was also Newton's view that gravity would be slightly weaker at the equator because it was farthest from the center of the Earth. This point was proved in the 1670s by several observers, when it was noticed that a pendulum clock, which kept good time in Europe, lost several minutes each day around the equator because the pendulum's acceleration due to gravity was weaker.

MEASURING THE EARTH

But could the overall shape of the Earth be established? The French academy of science organized two expeditions to measure two arcs of a <u>meridianal</u> degree, one in Lapland on the Arctic circle and the other on the equator in Peru. The Lapland party was led by Pierre Louis de Maupertuis, and the Peru expedition by Charles de la Condamine, both of them leading mathematical scientists. They left Paris in 1735, but the difficulties of travel with large surveying instruments were so great that the final reports from Peru were not announced in Paris until 1744.

By meticulous surveys from mountain tops both groups measured an exact arc of a meridian line running north–south and constructed its line of curvature by astronomical observations. When this arc was extended into a circle and divided by 360, the length of a single degree was arrived at. The Lapland figure was 69.04 miles, while that in Peru was only 68.32 miles. Thus the

Opposite: A plan of the solar system that includes the highly eccentric paths of comets; it was Halley who proved that the comets were an integral part of the solar system.

Below: Pierre Louis de Maupertuis, leader of the Lapland scientific expedition.

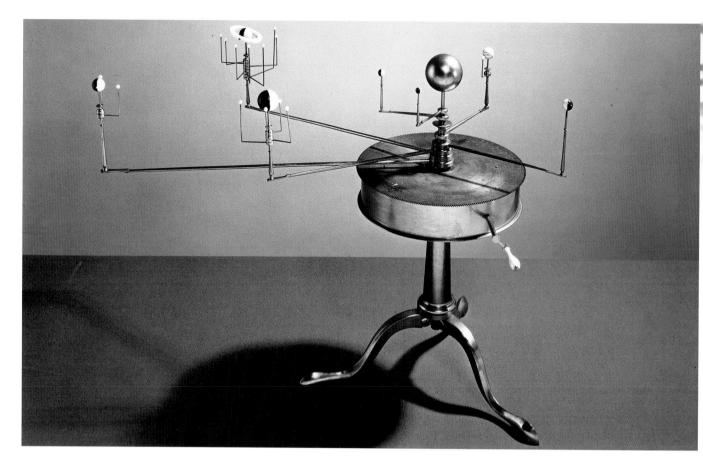

A small tabletop model of the solar system, including the moons of Jupiter and Saturn; the scale was impossible to show, of course, but the best of these models were geared so that the planets' relative periods of rotation were correct.

Earth's shape must be a sphere slightly flattened at the poles, as Newton's theory of gravity had predicted.

VARIATIONS ON A THEME

Another, and more profound, problem thrown up by Newton's theory concerned the dynamics of the heavenly bodies. In Newton's science the solar system was a vast, complex mechanism moving in obedience to unchanging laws. And yet astronomers had observed certain <u>anomalies</u> in those movements. Over a period of years the paths and speeds of the planets showed minute variations from those predicted by Newton's theory. The orbit of Jupiter, for example, appeared to be steadily shrinking, while Saturn's orbit seemed to be expanding. Most puzzling of all, the velocity of the Moon appeared to be steadily slowing by a fractional amount, making a month longer.

Newton himself had been aware of some of these anomalies, and he had suggested that although the universe was generally stable, it was still necessary for God to intervene from time to time to maintain its perfect equilibrium. This argument was welcomed by religious thinkers because it showed that the universe was not a self-regulating mechanism that could function without God.

A succession of great mathematicians, most of them French, set themselves to analyze these problems, and some of them concluded that the ether, the mysterious invisible substance that they believed filled the universe, created a dragging effect that disturbed the movements of the planets.

BALANCING OUT

The true solution, however, was worked out by Pierre Simon Laplace (1749–1827), who was able to show that the anomalies in the planetary movements were cyclical and periodic, and that they balanced each other out. For example, a slight decrease in the Moon's velocity was explicable as a result of a slight increase in eccentricity of Earth's orbit, and this effect was reversible. Laplace devised a series of equations to show that the sum of all the eccentricities of the planets is invariable, and that it is distributed in proportion to the planets' masses.

In effect, there is a fund of eccentricity in the solar system that is constant. If the eccentricity of one body increases, that of another will diminish, and these variations are all cyclical. The Jupiter–Saturn effect, for example, occupies a cycle of around 900 years.

These complex movements were all set out by Laplace in precise mathematical terms. The solar system was, after all, a highly stable system, and it was not necessary for God to intervene to balance it. Newton's theory had been vindicated in a way that was more precise than even he could have envisaged. Between 1740 and 1790 these mathematical advances served to establish the Newtonian view of the universe beyond all doubt. The rival Cartesian theory of the vortices was gradually forgotten, for the vortices could never be analyzed in this precise mathematical way; indeed, their very existence was purely hypothetical.

Pierre Simon Laplace, who analyzed the movements of the planets in enormous mathematical detail.

SCIENTIFIC LAWS

Pierre Simon Laplace (1749–1827) was the most important mathematical astronomer since Newton. He was the son a small farmer in Normandy, whose genius for calculation gained him a place in a military academy, and he began publishing mathematical papers before he was 20. His great work, *Treatise on Celestial Mechanics,* appeared in five volumes between 1798 and 1827, and it is one of the most technical, detailed, and forbidding works of science ever published.

Laplace was a leading figure in the French Enlightenment, whose work found special favor with Emperor Napoleon. Laplace believed that "All the effects of nature are only mathematical results of a small number of immutable laws," and his form of science clearly led toward atheism. In an interview with Napoleon the emperor pointed out to Laplace that in all his studies of the physical universe, he had never once mentioned the creator of it. Laplace famously replied, "I had no need of that hypothesis."

Laplace was also the author of the nebular theory of the origin of the solar system—that the Sun had condensed from a rotating mass of gaseous particles that then threw off the planets and their satellites. In one form or another this theory has prevailed to this day. Laplace's dying words were, "What we know is very slight; what we do not know is immense."

PIERRE LAPLACE (1749–1827)
- Astronomer and celestial mathematician.
- Born Beaumont-en-Auge, France.
- Studied at Caen, then Paris.
- Became professor of mathematics at the École Militaire.
- Researched the likelihood of Jupiter possessing satellites and the irregularities in the orbits of Jupiter and Saturn.
- Applied mathematics to physical astronomy, in particular analyzing the stability of orbits in the solar system.
- 1796 Published *Système du monde* ("The System of the World")—a collection of his ideas and theories on astronomy; his nebular hypothesis concerning planetary origins was added as a supplement to later editions.
- 1799 Joined the French Senate; 1803 became vice-president.
- Studied probability theory and the gravitational attraction of spheroids.
- 1799–1825 Publication in five volumes of *Mécanique céleste* ("Treatise on Celestial Mechanics"), a comprehensive explanation of celestial mechanics using applied mathematics.
- 1817 Louis XVIII made him a marquis.

William Derham and His Astro-Theology

A multiplicity of worlds. The realization that the Sun was a star led to the belief that there might be hundreds or thousands of solar systems, with planets like ours, perhaps inhabited by creatures like us. This multiplicity of world's was thought to display God's omnipotent power (see panel opposite).

It might be thought that the new science, with its aim of observing and deducing natural laws, might come into conflict with the religious view that the universe was governed solely by God. In fact, the eighteenth century saw the growth of a novel form of "natural theology" in which the complexity and harmony of the universe were used to prove the existence and the power of God.

One of the most influential texts of natural theology was a work called *Astro-Theology, or a Demonstration of the Being and Attributes of God from a Survey of the Heavens.* The author, William Derham (1657–1735), was a parish priest and an amateur scientist of no great genius, but he was in touch with contemporary ideas. His book obviously struck a chord, because, after its first appearance in 1715, it was reprinted 10 times in the next 50 years.

Derham explained that he had made considerable use of a telescope built by Huygens. The subject of his book was the implications for religious belief of what the telescope had revealed about the physical universe. The central point at issue was the scale of the universe.

Since Copernicus's theory first began to be accepted, it had been evident that the idea of a finite universe might no longer be tenable, and the telescopic age had gradually convinced astronomers that the Sun was simply a star, and that the stars were all suns in their own right. This had enormous implications for ideas about the scale of the universe and about the Earth's place in it.

In traditional religious thought the Earth and humankind were God's special creation, the center of the universe, but this was now shown to be scientifically untrue. Derham, however, was not troubled by this, affirming instead that "the telescope has laid open a new and far more grand and noble scene of God's work than the world before dreamed of."

THE THIRD WAY

He described the two leading theories of the structure of the universe—the Ptolemaic and the Copernican—but said that they had been replaced by a third system in which the sphere of the fixed stars was no longer believed to be the boundary of the universe at a fixed distance from the Sun. "The new system," wrote Derham, "supposes that there are many other systems of suns and planets, namely that every fixed star is a sun, and encompassed with a system of planets, as well as ours."

How did Derham answer the charge that this removes humans from their unique position in creation? He argued that "This is far the most magnificent of any system, and worthy of an infinite creator, whose power and wisdom are without bounds and measure, and so may in all probability exert themselves in the creation of many systems."

The idea of a "plurality of worlds" is more to the glory of God than a single world. Although he could not see the planets with the telescopes then available, Derham was convinced that they existed, and moreover, he believed they were inhabited, for why should God create so many stars except to act as suns to give life to other worlds as ours does to the Earth?

VAST AND ORDERLY SYSTEM

What was the scale of this universe? Derham was unsure, but he made some calculations that showed it to be far greater than was formerly believed; his argument depended on the idea that all the stars are more or less equal in size, so that their brightness varies only because of their distance from Earth:

"Forasmuch as few stars appear other than as points, even through our best telescopes, how prodigiously much further they must needs be from us than the Sun is, to cause their appearance to be so very much less than the Sun.
For example, let us take one of the fixed stars supposed to be nearest to us, as being the brightest and largest, namely Sirius. Now this, by the accurate observation (of Huygens) hath been found to be in appearance 27,664 times less than the Sun, and consequently by the foregoing rule, it is so many times further off than the Sun is, which will amount to above two millions of millions of English miles, and if so what an immeasurable space the firmament is."

Derham's central argument was that as science progresses and reveals more and more of the immensity and complexity of the universe, it proves that God's powers are even greater than previously believed. There need be no conflict between science and religious belief. Everything in the universe is harmonious and ordered, and reveals the character of its creator.

This is the argument known as teleology—that something that is designed and that functions harmoniously must have a designer. The universe of eighteenth-century astronomy was a vast and orderly system, and in Derham's thought people's growing understanding of it could only bring them to a deeper reverence for its creator. His text is entirely typical of the rational or natural theology that flourished in Britain at the time.

Thomas Wright in 1741, by Thomas Frye (1710–62).

A "MULTIPLICITY OF WORLDS"

This idea became popular in the eighteenth century. Thomas Wright (1711–86) suggested that massive star groups, such as our own galaxy, existed throughout the universe, and that many stars possessed satellites like the earth. He believed that God had created this "multiplicity of worlds," and that people might be reincarnated in other worlds. Wright argued that theology was nothing without astronomy— that we must study the structure of the universe in order to understand God. He believed that the universe had a large scale structure and a "divine center" where God was located, and that successive reincarnations would take us closer to this center. Wright accepted completely Derham's claim that the vast universe revealed by post-Copernican astronomy displayed God's greatness and creative power.

Pioneers of Scientific Botany

The mystery of life resolved itself into the twin problems of form and function: studying and charting the separate physical structures that made up the animal or the plant, and then devising experimental means to discover what role they performed. The plant was seen as a simpler life form than the animal in many ways and more amenable to experiment, making botany an attractive field for scientists.

The pioneer of plant physiology was Stephen Hales (1671–1761), one of the many English clergymen whose chief energies were spent not in the study of theology but in studying the world of plants, birds, or insects. Hales was unusual in going beyond descriptive botany into experiment and measurement. His chief interests were in plant nutrition, and whether there was any equivalent in plants to the circulation of the blood that had been discovered in animals.

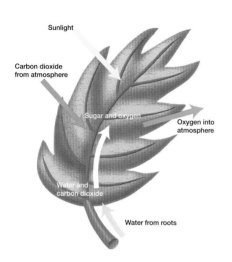

Sunlight

Carbon dioxide from atmosphere

Sugar and oxygen

Oxygen into atmosphere

Water and carbon dioxide

Water from roots

Photosynthesis—several botanists were already convinced that plants draw part of their nourishment through their leaves from the air, but it was Ingenhousz who first analyzed the fundamental process of photosynthesis. Plants convert water and carbon dioxide into the sugar with which they grow, and they produce oxygen as a by-product.

EXPERIMENTS WITH PLANTS

It was common knowledge that sap rises through the plant, but what did the sap do, and did it rise and then fall in a cycle? Hales first communicated some of his ideas to the Royal Society in London in 1718 and then spent a further eight years in experiments. His major work, *Vegetable Staticks,* was published in 1727. In this book Hales described his method of testing plant <u>physiology</u> by cutting the limbs of plants and inserting a simple pressure gauge. He discovered that the sap rises with a measurable force that varies at different seasons of the year and is influenced by temperature. He showed that branches will take in water not just from the roots, but from the upper ends of a cut limb—in other words, the flow direction is reversed from the normal—and he correctly saw that this meant there could be no circulation within the plant like that of an animal.

THE ROLE OF THE LEAF

His greatest discovery was the part played by leaves in plant nutrition. He measured growth through the seasons and found that growth rates are fastest when plants bear leaves. He was therefore convinced that "plants draw through their leaves some part of their nourishment from the air," and that the function of the leaves was "to perform in some measure the same office for the support of vegetable life that the lungs of animals do for the support of animal life." He also speculated that light might play a part in plant growth, but this he could not prove. Hales had demonstrated that plant nutrition was more complex that had been imagined. Although they absorbed moisture from the soil, plants did not simply receive their nutrition in this way; respiration through the leaves was also involved.

CIRCLE OF LIFE

It was the Dutch scientist Jan Ingenhousz (1730–99) who clarified what we now call <u>photosynthesis</u>. Ingenhousz was helped by the fact that chemistry had made great progress in the 1770s, and that it was clearly understood that animal respiration used one constituent of the air and replaced it with another constituent. He performed experiments that proved that "plants purify the common air in the sunshine and injure it in the shade and at night." In other words, they absorb carbon dioxide and emit oxygen—but only the green leaves can do this, and then only in the presence of sunlight.

Ingenhousz proved this by enclosing plants in containers and then collecting and analyzing the gases at different times and under different conditions. From these results he concluded that plant and animal life are linked in a cycle, or an "economy" as it was called in the eighteenth century. Perhaps Ingenhousz deserves to be called the first scientific ecologist.

HOW PLANTS GERMINATE

A further example of this link between plants and animals was beautifully demonstrated by the German naturalist Christian Konrad Sprengel (1750–1816), whose chief interest was the study of plant germination. It had been understood for some time that many plants were hermaphrodite, that is, of both sexes, and it had been assumed, therefore, that they fertilized themselves. It was Sprengel, however, who pointed out the crucial fact that the male and female parts of the plants, the <u>stamens</u> and the <u>anthers</u>, mature at different times, and therefore self-fertilization is impossible. Instead, he drew attention to insect visits, noting that color and scent are attractions, and observing how <u>corolla</u> markings guide the insects to the hidden nectar. He concluded that insects cross-pollinate flowers, and that "Nature appears not to have intended that any flower should be fertilized by its own pollen."

Much later, Charles Darwin studied Sprengel's work and went on to show how indispensable cross-pollination was to plant evolution: No new mutation could occur, and no new species could evolve if plants fertilized themselves. Thus the insect–plant relationship so elegantly described by Sprengel may be seen as the first attempt to explain the origin of organic forms from a definite relationship to the environment. Christian Sprengel can thus be seen as contributing some of the very earliest insights both of <u>ecology</u> and of evolution.

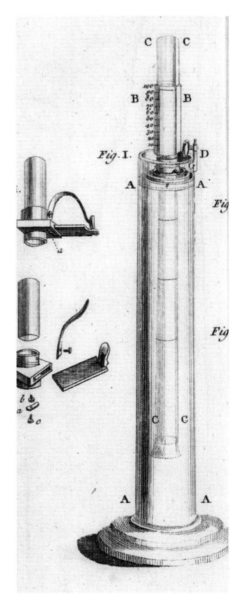

Equipment used by Jan Ingenhousz —the frontispiece of the 1779 *Experiments upon Vegetables,* **discovering their great power of purifying the common air in the sunshine and of injuring it in the shade and the night. Ingenhousz conducted over 500 experiments and wrote up his findings in the book.**

The Argument about Spontaneous Generation

It may seem strange that in the middle of the Age of Enlightenment, when scientific botany and zoology were becoming established, the old controversy about whether life could be spontaneously generated should be revived. There was, however, a good philosophical reason for it. The <u>rationalist</u> thinkers in France were seeking an alternative to the traditional belief that all species of life had been created by God's divine decree. Therefore, they seized on any evidence that, given the right conditions, life could arise spontaneously. For this reason the leading French naturalist of the age, Comte de Buffon, gave his authority to the idea in spite of the fact that Francesco Redi had disproved it in the 1660s (see Volume 5, page 56).

CHANGE AND GROWTH

Buffon's immediate source was some experimental work carried out by an English scientist who lived mainly in France, John Needham (1713–81). In 1748 Needham believed that he had proved that spontaneous generation occurred at the microscopic level in liquids containing either plant or animal material. Needham boiled mutton and then sealed the juices in airtight containers. Some days later he observed small organisms moving in the liquid. Moreover, he convinced himself that they grew and changed shape. Buffon accepted the validity of these experiments and published his own conclusions, thus in the process reviving the whole controversy. Buffon suggested that living tissue decomposed into organic building blocks that spontaneously reconstituted themselves into different forms of life.

A NATURAL EXPLANATION

The scientist who set out to test whether spontaneous generation was fact or fable was a compatriot of Redi, Lazzaro Spallanzani (1729–99), who was, for many years, professor of natural history at the University of Pavia in Italy and a great experimentalist. His motto in scientific work was to have no preconceptions but simply to "interrogate nature"—so he tried one method after another to completely isolate the same type of organic liquids that Needham had used from the outside environment. Only in this way could he establish whether the new growth was truly spontaneous or had a natural explanation.

Comte de Buffon developed a rational, nonbiblical approach to the study of nature, but many of his ideas proved to be quite wrong.

After a great deal of trial and error he found that the liquids remained free of all microorganisms when they were put into flasks that were hermetically sealed and the contents boiled for an hour. If air was let in, organisms did grow after a certain time. Spallanzani was convinced that Needham had simply not sterilized his containers correctly, and that the organisms that Needham had seen were the natural offspring of organisms that existed in the air. Spallanzani's work was conclusive; but because it was written in Italian, it was not widely known, and the idea that he had disproved did not vanish at once.

PREFORMATION DISPROVED

The mechanisms of generation were still not well understood, and the discovery by the Swiss naturalist Charles Bonnet (1720–93) that female aphids reproduced without contact with the males raised once again the idea of preformation, that the actual form of successive generations of creatures was contained in the eggs. This belief was welcomed by religious thinkers, who took it to prove that God had created the form and substance of all future generations in one creative act. The experiments that finally disproved preformation were performed by the German naturalist Caspar Friedrich Wolff while he was working in St. Petersburg as a member of the scientific academy there. In detailed studies of chick embryos Wolff showed clearly that the heart and blood vessels and other internal organs all grew and became differentiated out of a formless liquid. Wolff's works became widely known, and the old idea of preformation was soon consigned to history, while scientific embryology became of central importance.

The metamorphosis of butterflies and moths seemed a clear argument against preformation, yet some naturalists claimed that the wings were already formed inside the caterpillar. It would take the work of Friedrich Wolff to dismiss this idea.

Rational Theories of Nature:
Buffon and Lamarck

JEAN-BAPTISTE LAMARCK (1744–1829)
- Evolutionist and naturalist.
- Born Bazentin, France.
- Served in the French army.
- While working in a Paris bank, started to study medicine and botany, in particular Mediterranean flora.
- 1773 Published *Flore française* ("French Flora"), a popular success.
- 1774 Appointed keeper of the royal garden.
- 1793 Became keeper of invertebrates at the new Natural History Museum; his job included lecturing on zoology.
- Started working on the taxonomic distinction between vertebrates and invertebrates.
- By about 1801 his work led him to thinking about the relationship between species and their origins.
- 1809 Published *Philosophie zoologique* ("Zoological Philosophy") in two volumes, containing his thoughts and conclusions on evolution—that species need to adapt to cope with the changes in their environment, and that certain attributes are passed down the generations.
- 1815-22 *Histoire naturelle des animaux sans vertèbres* ("Natural History of Invertebrates"), his major work on invertebrates.

The most decisive process of rethinking in the life sciences occurred in France, where Buffon and Lamarck set out to give accounts of nature's workings without reference to God. Both men saw nature as in a state of flux, of gradual change and development over long periods of time—far longer than that implied by the traditional biblical time frame. The crucial point for both men was the abandonment of the doctrine of the fixity of species, the belief that all life forms were created by God at one moment. Instead, they affirmed that new life could arise, and that species could change and develop over time. In this sense they are important precursors in the history of evolutionary thought.

THE BIRTH OF PLANETS

Georges Leclerc, Comte de Buffon (1707–88), was an aristocrat who employed his wealth and his country estate in the study of natural history. One of his most original theories concerned the origin of the solar system. He proposed that a comet had struck the Sun and torn off masses of hot liquid and gas that had then solidified into the planets and their satellites. The Earth had steadily cooled, allowing rocks and seas to form. Buffon estimated that this event must have occurred more than 75,000 years ago, compared with the traditional age of the Earth of 6,000 years that scholars had deduced from the Bible. (There is some evidence that Buffon's private calculations suggested a time frame even longer—millions of years—but that he felt this figure would not be believed and so reduced it.)

The comet theory was attractive because it seemed to explain why all the planets should be moving in the same direction in a common plane. Its weakness was its assumption that comets were themselves dense stars, massive enough to have this effect on the Sun; in reality a comet would be consumed long before it touched the Sun's surface.

LIFE ON EARTH

After the birth of the planet, Buffon affirmed that life had emerged spontaneously from the warm earth: No divine act of creation was involved. This event was not unique, but occurred as many times as there are species, for Buffon did not believe that the great classes of animals—mammals, reptiles, fish, birds, insects—were related to each other in any way. Indeed, even within a class, for example, the mammals, he denied that horses, lions, dogs, elephants, and so on were mutually connected. He did allow that species within a family, say, horses and asses or lions and leopards, were probably related through a common ancestor. But in Buffon's thought the relationship was one of degeneration, not positive development. The ass might be a degenerate horse, or the ape might be a degenerate human.

This idea sounds quaint to us, but it sprang from Buffon's belief in spontaneous generation. The organisms that formed as the Earth cooled developed spontaneously in many different ways, thus giving rise to the many distinct families of animals and plants. Thereafter, the individual species became transformed in response to their environment.

Buffon's approach to the biological sciences was, perhaps, vague and theoretical, and it cannot be claimed that he was on the verge of understanding evolution. But he did see nature as a creative power and life as participating in a long process of change. In this he was very influential, especially in France, where a rational, nonreligious basis of science was the ideal.

CHANGING LIFE

The naturalist who went even further than Buffon in rejecting the idea of the fixity of species was Jean-Baptiste de Lamarck (1744–1829). Like Buffon, Lamarck believed that the Earth and its life forms had developed over tens of thousands of years. "Time," he wrote, "is always at the disposal of nature and represents an unlimited power with which she accomplishes her greatest and smallest tasks."

Where Lamarck differed from Buffon was in his belief that nature had produced only a few original organisms, and that all the species that we now see had developed from those few. Lamarck was convinced that life could and did move toward increasing complexity of form. In this movement the influence of the environment was paramount: Changing circumstances and changing physical needs evoked a direct response, so that animals became better adapted to the environment. The classic example that Lamarck gave was the long neck of the giraffe—as the animal stretched after food growing high up, its neck became longer.

Central to Lamarck's theory was that these newly acquired physical characteristics would be inherited. We now know that this cannot happen in one or two generations, any more than the skills of a great artist or athlete will necessarily be inherited by his children. Lamarck imagined a very direct sequence of environmental challenge followed by a biological response. In this he was mistaken, but there is no doubt that his work established the idea that nature's forms might be evolving and becoming more complex.

Not everyone agreed with Lamarck about how classes and families of animals were related to each other. His suggestion that the development of humans themselves might be seen as the result of the same processes was still controversial even in France. Taken together, the ideas of Buffon and Lamarck aimed at establishing biology on a new footing as an autonomous science free of theological influences.

Above: Lamarck's theory that animals evolved to a complexity of form is perfectly exemplified by the giraffe and its long neck.

Below: The introductory page to Buffon's great work.

The Science of Taxonomy: Carl Linnaeus

Opposite: Linnaeus's system of classifying plants according to the number and arrangement of their anthers and stamens; the 24 primary classes thus distinguished are shown here.

The large-scale theorizing about nature from writers such as Buffon and Lamarck represents one aspect of eighteenth-century life science. But at the same time, naturalists were engaged in a meticulous process of naming and trying to classify all the known species. The outstanding figure in this science—called taxonomy— was the Swedish scholar Carl Linnaeus (1707–78), whose major achievement was in the field of botany.

Trained as a physician, Linnaeus was attracted to botany, and from a very early age he became obsessed with the aim of reforming plant classification. He sought some intellectual key that would enable him to group plants according to their structure and type. That was becoming more important since innumerable unknown plants were being collected and brought back to Europe by travelers in America and Asia. Botanists felt that their science was being overwhelmed by new specimens whose relationship to known species was totally unclear.

FINDING THE KEY

In 1730, when Linnaeus was just 23 years old and a student at the University of Uppsala, he discovered the key that he had been seeking in the sexuality of plants. He began to make notes and observations that over the following five years, would grow into his new system of plant classification, and that would begin a revolution in the language of botany.

Linnaeus saw that the whole plant kingdom could be grouped according to the structure of the male and female sexual parts of the flower, the stamens and the pistils. The first step was to create large groups of plants that had one stamen, or two, or three, or more; this group Linnaeus called the class, and he named them monandria, diandria, triandria, and so on.

Then the botanist should look at the female pistils, and within each class he or she should further subdivide according to their number; this subdivision was called the order. Finally, within each order came the type, or genus, and then the species.

It was in identifying genus and species that Linnaeus introduced his most celebrated innovation—the binomial system in which two Latin words give a unique label to any plant. Thus goat-willow is *Salix caprea,* where *salix* is the Latin for willow, and *caprea* means goat; mistletoe is *Viscum album,* meaning white mistletoe. In each case the second word offers some description of the plant—it may be

The title page of Linnaeus's *Systema Naturae* ("Nature's Systems").

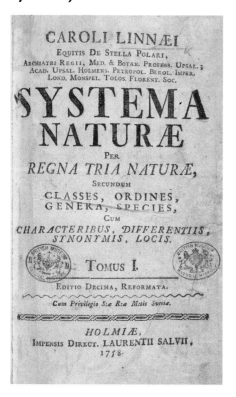

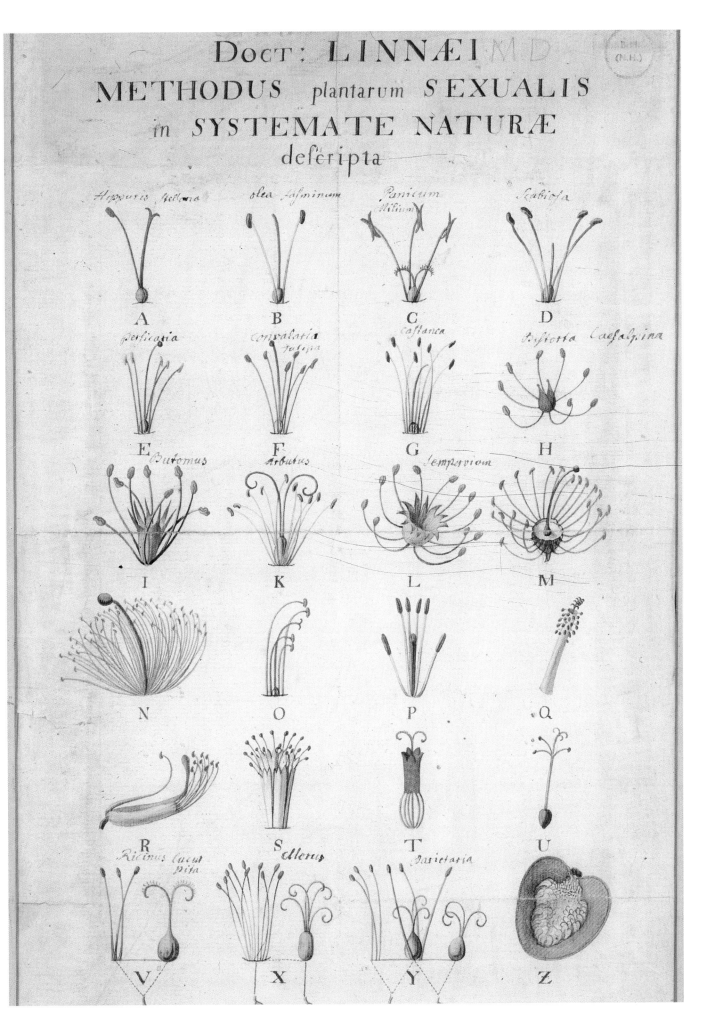

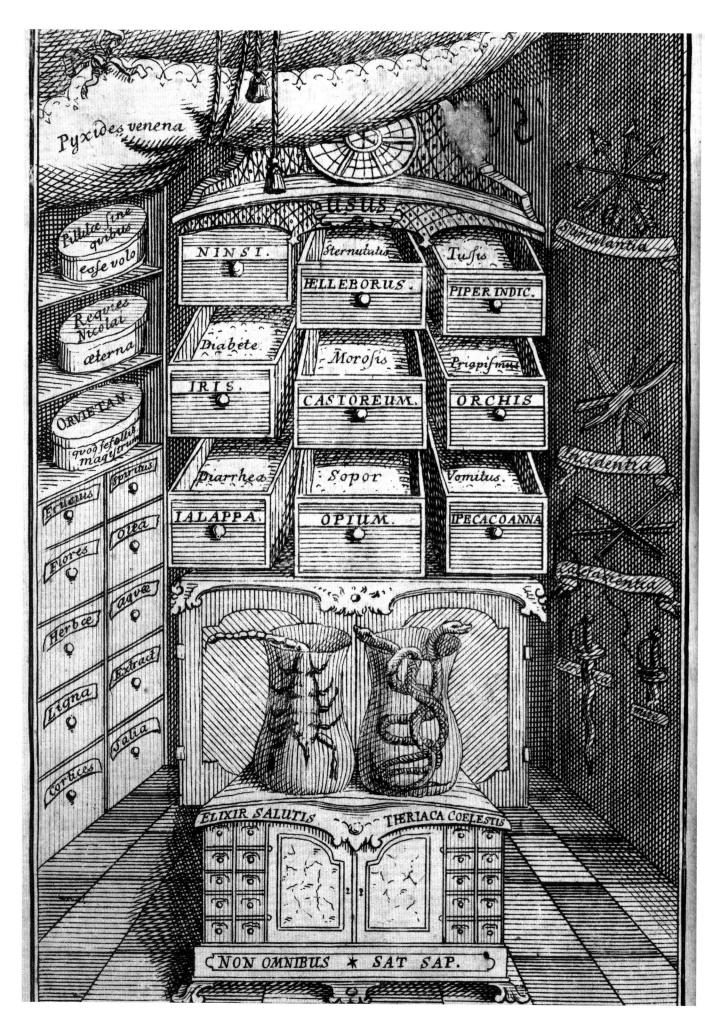

the color, size, or some other association —for example, the name of an animal that eats it.

PRACTICAL TOOL

This system was so comprehensive and flexible that any new plant could quickly be grouped, named, and identified in a way agreed to by all other botanists. Some critics objected—and Linnaeus realized this—that it was an *artificial* system. It selected one particular characteristic of a plant and based its classification on that alone. It ignored other aspects of plant structure, and it suggested connections between plants that might be purely external or accidental. Linnaeus offered no scientific definition of the terms class, order, genus, or species, but his system worked as a purely practical tool. In his day Linnaeus was regarded as a genius, for he seemed to have grasped the order and harmony of nature in a way that some admirers felt was almost as important as Newton's.

Linnaeus was not a great experimentalist, and he made no contribution to plant physiology. He was a conservative, religious man, who had no time for the speculation of Buffon and others. He regarded himself as an intellectual spectator in the vast garden that God had created. He once wrote: "I saw the infinite, all-knowing and all-powerful God from behind as he went away, and I grew dizzy. I followed his footsteps over nature's fields and saw everywhere an eternal wisdom and power, an inscrutable perfection."

Above: Linnaeus, the great classifier of the natural world.

Opposite: The greatest practical application of botany was in pharmacology – the preparation of natural drugs from plants – which Linnaeus described: each drawer contains extracts from plants such as hellebore, poppy, ipecacuanha, orchid, and so on.

Mechanism versus Animism: Georg Stahl

With the understanding of the circulation of the blood, and with the triumph of mechanics in physical science, the physicians of the early eighteenth century felt that they were on the verge of discovering the secret of how the human body functions. Harvey had proved that the heart was a pump, and the veins and arteries were the pipes that led to and from it. Borelli had analyzed the movements of the spine and limbs in terms of levers and pulleys. It was only a matter of time, the physicians felt, before all the workings of the human body could be reduced to mechanical cause and effect.

The chief advocate of this school was Hermann Boerhaave (1668–1738), teacher of medicine at the University of Leiden in Holland, who influenced a whole generation of physicians and philosophers. Boerhaave taught that all the processes in the human body worked through physical interactions or pressures. Particles or fluids were sent from one part of the body to another to cause functions such as respiration, nutrition, sensation, or movement. These particles or fluids often moved through channels invisible to the scientist—for example, the fluid that was thought to fill the nerves and convey messages. When these particles or fluids became congested or imbalanced, sickness would result. It was the quantity, weight, and pressure in these systems that were crucial to health and sickness.

THE HUMAN MACHINE

It is difficult not to see these systems as a new version of the ancient theory of the four humors (see Volume 2, page 58), except that now they have a more scientific appearance, being tied to specific systems such as the digestion or the nerves. This mechanistic model of the human organism flourished because there was no knowledge of the chemistry of the human body or of the cell as the functional unit. Eighteenth-century physicians such as Boerhaave were seeking a new key to physiology, a new organizing principle, but they were moving on largely by guesswork. They believed that it would be the rules of physics, and not any other science, that would explain physiology.

This mechanistic view of the human organism was given its clearest expression by the French physician and philosopher, Julien La Mettrie (1709–51), who had been a pupil of Boerhaave. La Mettrie was an outright atheist whose two major works indicate the character of his thought. He wrote *The Natural History of Soul* in 1745 to show that all the psychological or spiritual states of the human mind are really caused by physical changes in the brain and

HERMANN BOERHAAVE (1668–1738)
- Physician and botanist.
- Born Voorhout, Netherlands.
- 1682 Studied theology and oriental languages at Leyden, then in 1689 took his degree in philosophy.
- 1690 Started studying medicine and was lecturing on the theory of medicine by 1701.
- 1708 Published *Institutiones Medicae* ("Medical Principles") and the following year *Aphorismi de Cognoscendis et Curandis Morbis* ("Aphorisms on the Recognition and Treatment of Diseases"); both were widely translated and won him international medical recognition.
- 1709 Appointed professor of medicine and botany.
- His theorizing led him to the belief that plants and animals show the same law of generation.
- 1718 He was teaching about sexual reproduction in plants.
- 1724 Appointed professor of chemistry and published *Elementa Chemiae* ("Elements of Chemistry").
- His fame enticed patients from all over Europe to consult him, and he became very wealthy.

body and then *The Human Machine* in 1748 to develop Boerhaave's model of the human body as a mechanism that worked through the interaction of physical parts.

La Mettrie taught a purely materialistic view of human nature: Only science could offer any genuine knowledge of humankind. Mind had a physical basis, and the soul was a myth, for what had been called the mind or the soul actually worked by a superior form of physics that scientists would one day discover. Put crudely, a human was "a machine that winds its own springs." He satirized the medical profession for its backwardness and regarded religion as a superstitious survival. Even in an age of enlightenment La Mettrie's ideas were so radical that he often had to move hastily around Europe to escape the enemies that he made.

Intellectually, the first weakness of La Mettrie's position was that in his day there were really no machines that were complex or subtle enough for him to point to as a models for the complexities of the human body. Second, there was the perennial problem that faced any mechanistic philosophy. How could the mind or the will or the consciousness interact with the body? What directed all the movements of the human machine?

The Physician—a 1760s engraving by J. Falkema after a painting of 1659 by C. Netscher.

THE VITAL PRINCIPLE

Those scientists who felt the force of this last question set out to develop a theory of human nature that was opposed to mechanism, and that is known as vitalism or animism. The principle of this school of thought was that life itself is some mysterious force outside the physical body, and that this manifests itself as mind or spirit in each individual. The Swiss physician Albrecht von Haller (1708-77) was an influential member of this school of thought. He had made a special study of twins born joined at the abdomen who shared one physical system for the performance of their vital functions, but who plainly had two wills and two personalities.

As La Mettrie had stated mechanism in its extreme form, so the exponent of animism in its classic form was the German Georg Stahl (1660–1734). Stahl's starting point was his emphasis on the irreducible difference between the living and the nonliving. Stahl challenged the mechanists to define exactly where the difference lay between a living body and corpse; no mere physical description of bodily systems could account for the fundamental difference. So Stahl assumed the existence of a vital principle, a controlling agent, which he called the *anima*, from the Latin word for soul. In Stahl's view, however, it was not a religious or mystical concept, but a scientific principle.

GEORG STAHL (1660–1734)
- Chemist.
- Born Ansbach, Germany.
- 1694 Became professor of medicine at Halle.
- 1714 Appointed personal physician to the king of Prussia.
- His phlogiston theory of combustion and fermentation.
- Developed the theory of animism.

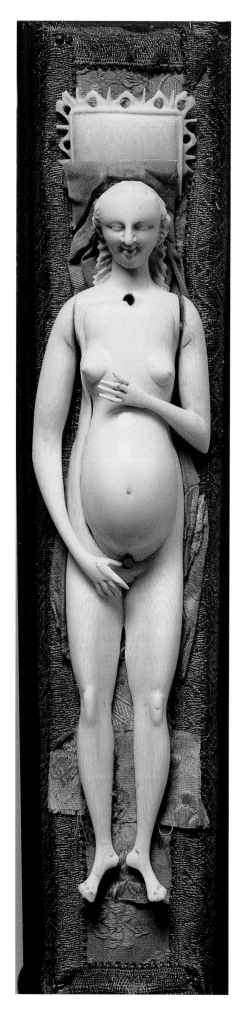

A PURPOSEFUL LIFE

Stahl pointed to another weakness of the mechanistic view—that it fails to explain how and why human actions and life are directed or are purposeful. Human activity does not consist only of physical movement, but seeks goals that are social or intellectual, of which the mechanists could give no explanation. He gave the example of the apparently simple act of leaping a ditch. The desire to leap is conceived by the anima, and then all the individual movements of muscles and bones in arms and legs follow from that desire without conscious control. The model of the machine cannot explain this process of goal achievement. This example also shows that the anima was not simply equivalent to consciousness. The anima was involved in two levels of activity—both in automatic physical processes such as digestion or vision, but also in acts of will. The anima was also implicated in disease, which was the soul's attempt to expel morbid matter and reestablish bodily order.

Ultimately Stahl could not explain what the anima was, but he had clearly exposed some of the weaknesses of the mechanistic school. He had shown that in the current state of science, a physical description of the human organism was simply impossible. Moreover, there were good philosophical reasons for thinking that some vital principle existed in living organisms that would never be analyzed into mechanical forces. The balance between mechanism and animism in the life sciences would seesaw back and forth over many years. In the longer term the mechanistic view would prevail, but in ways that neither Boerhaave nor La Mettrie could have foreseen, for physics alone would not provide the answers.

PHLOGISTON

Georg Stahl was also responsible for one of the most influential ideas in eighteenth-century physical science: the theory of phlogiston. For centuries alchemists and craftsmen knew that fire is the most powerful agent for changing physical substances, either combining, separating, or destroying them. But what fire really was could never be explained in chemical terms. Before Stahl many alchemists believed that all matter consisted of several principles that made it solid, liquid, or burnable; this was an idea that had developed from the earlier doctrine of the four elements. On the basis of this idea Stahl built his theory that when matter burns, a certain constituent is being liberated; this constituent he named phlogiston, from the Greek word meaning "burned."

He believed that phlogiston was present to some extent in all matter, but that substances such as wood contained much more of it than, for example, iron. Stahl realized that combustion only takes place in the presence of air, but he argued that the function of the air was simply to carry away the phlogiston; it was not the air itself that was involved in the burning, and flames were the phlogiston whirling away into air.

When metals were heated fiercely, they were reduced to a metallic ash called calx, from which it was deduced that metals were composed of calx plus phlogiston. If the calx were re-heated over charcoal, the process would be reversed, as the metal took up the phlogiston from the charcoal and returned to its metallic form. The idea of phlogiston offered an attractive explanation of one of the most fundamental and common of all chemical transformations. Weight was given to Stahl's idea because it came from eastern Germany, which had long been a center of mining and metallurgy, and his chemical works were translated into many European languages, so that phlogiston soon became scientific <u>orthodoxy</u>.

WEIGHTY PROBLEMS

However, there were certain difficulties with the theory. First, many chemists knew that when some substances are burned, the remaining ash or calx could actually be heavier than the original matter, which was hard to reconcile with the idea of phlogiston being released. Stahl answered this objection by arguing that phlogiston was a highly volatile principle that had "negative weight," or buoyancy, so that when it was given up, the substance became heavier.

The other problem was that experiments strongly suggested that burning actually used up some part of the air: A candle in a bell jar soon goes out, but what is left in the bell jar is not vacuum, but a further kind of gas or air. Stahl knew this, but believed that the air could absorb only so much phlogiston before it became saturated with it, and at that point the burning ceased. Stahl called that air "phlogisticated air," and we would call it carbon dioxide.

However, one of the most intriguing details of Stahl's system appeared when he faced the question: What happens to all the phlogiston that burning liberates into the air? He knew that wood and charcoal burn well, therefore he reasoned that plants were able to absorb "phlogisticated air," which is exactly what they do. Thus Stahl was led to theorize about the necessity of a carbon cycle more than 100 years before that cycle began to be properly understood.

Stahl was an original thinker who, in an age of experimentation and mechanical philosophy, still felt intuitively that matter was directed by subtle principles, and he built his intuitions into a form of science. He once wrote, "Where there is doubt, whatever the great majority of people believe is wrong."

Opposite: Eighteenth-century anatomical figure of a pregnant woman. The figure had removable organs and was used for obstetrics training.

Stahl's phlogiston theory supposed that all burnable matter contained a subtle substance that was released during burning. Stahl recognized that wood is among the most flammable of materials, and he suggested that plants absorbed phlogiston as they grew, which was then released again during burning. This can be seen as anticipating the idea of the carbon cycle.

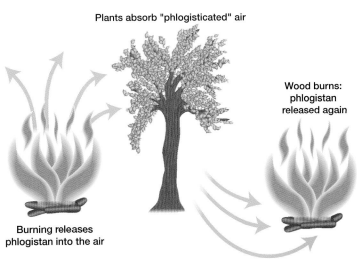

Plants absorb "phlogisticated" air

Wood burns: phlogistan released again

Burning releases phlogistan into the air

Mapping the Human Form

ALBRECHT VON HALLER (1708–77)
- Anatomist, physiologist, and botanist.
- Born Bern, Switzerland.
- Studied anatomy and botany at Tübingen and Leyden.
- 1729 Started in medical practice.
- 1736 Appointed professor of anatomy, surgery, and medicine at the new University of Göttingen.
- At Göttingen started an anatomical museum and theater. Helped found the Academy of Sciences and organized a botanical garden, among many other activities.
- Discovered the mechanical workings of the heart muscle.
- Investigated techniques of using injections.
- 1753 Returned to Bern to become a director of a saltworks and a city magistrate.
- Devoted much of his time to writing novels and scientific works.
- 1757 Published *Elementa physiologiae corporis humani* ("Physiological Elements of the Human Body") in eight volumes—an explanation of how the body functions.

Eighteenth-century physicians produced superbly detailed and accurate charts and models of human anatomy.

The philosophical arguments between mechanists and animists about the nature of life did not greatly influence the everyday practice of medicine. Even in the Age of Enlightenment death was everywhere, and the early years of the eighteenth century saw European cities still at the mercy of epidemic diseases, for worldwide travel encouraged the spread of killers such as bubonic plague, smallpox, and typhoid. Conditions in institutions such as hospitals and prisons, or in the army and navy, were still appallingly unhygienic, so that disease could thrive and spread through society.

The empirical approach to science had one outstanding effect on medical science, however, and that was to stimulate anatomical studies. The taboos on human dissection were steadily weakened, and physicians were able to study and map every aspect of the human body in greater and greater detail, seeking in the form of the bodily parts clues to their vital function. Improvements in techniques of drawing and printing resulted in the finest atlases of anatomy ever seen. The Swiss physiologist Albrecht von Haller (1708-77) of the University of Gottingen published his *Icones Anatomicae* ("Images of Anatomy") in several parts between 1743 and 1756, with large illustrations of all the regions of the human body laid open at dissection, showing organs, muscles, arteries, and nerves. These pictures were on such a large scale that Haller also provided outline key diagrams to identify the numerous features, running into many hundreds for each plate.

PRECISE DRAWINGS

Where Haller showed the regions of the body, his contemporary, Bernhard Albinus (1697–1770) of Leiden University in Germany, set himself to depict the entire human form. Years of study, preparation of the subjects, and cooperation with a superb Dutch artist named Jan Wandelaar resulted in the most precise and elegant anatomical plates ever published, in the atlas entitled *Tabulae sceleti et musculorum corporis humani* ("Tables of the Skeleton and Muscles of the Human Body"), 1747.

Albinus was driven by the desire for precision to record objectively the reality of the physical body. In his view this meant showing the bones and muscles as they are in life, standing erect and moving. In order to *portray* this reality for others to see, he was therefore compelled to make very complicated preparations. He had to mount the human bodies with wires and ropes, and treat them

Mesmerism: Patients surround a large vat that Mesmer claimed contained "magnetic fluid" that restored health.

with vinegar or other preservatives during the days or even weeks while Wandelaar was making his meticulous drawings. He found it best to do this outdoors in frosty weather, thus keeping the subjects firm and delaying putrefaction. When positioning the dead subject, he used for reference:

"a thin man of the same size as my skeleton, and making him stand naked in the same position, I compared the skeleton with him. However, the frozen state of the skeleton was disturbed by the fire that we were obliged to have always when the naked man stood, for he neither could nor would stand without it."

The artist ensured absolute accuracy of proportion by erecting a net of squared cords in front of the subject from which he carefully transferred every feature onto his squared paper. Albinus placed his subjects against natural backgrounds to emphasize their perfect proportions, and this gives his pictures a classical quality.

The purpose of Albinus's work was to provide the scientist with an absolutely faithful record of the perfect, healthy human body. From such knowledge the physician would be better able to detect any departure from the normal; in other words, systematic pathology could be developed.

DETAILED STUDY

It was the Italian Giovanni Morgagni (1682–1771) of the University of Padua who really founded the discipline of modern pathology. In 1761 Morgagni published *De sedibus et causis morborum per anatomen indagatis* ("On the Roots and Causes of Diseases Investigated by Anatomy"). This work describes the results of hundreds of postmortem investigations that Morgagni had carried out in which he had investigated in detail the differences between the healthy and unhealthy body. He announced the basic principle that every anatomical alteration in an organ or part of the body was accompanied by an alteration in function, and that this will often result in disease and death. The heart, the lungs, the brain, the eye, the joints—all were examined and their condition related to the symptoms exhibited before death. To give one outstanding example, Morgagni showed in cases of brain damage that the paralysis affected the opposite side of the body to the side of the brain that had been injured.

ANIMAL MAGNETISM

The mechanisms of physiology by which the human body performed its many complex functions still eluded scientists in the eighteenth century, yet the detailed research and observations of Haller, Albinus, Morgagni, and many others did signal the end of some traditional medical myths. The idea that sickness was a result of the ebb and flow of the four humors and was influenced by the stars and planets, so that astrology was an essential part of the physician's training—these doctrines were at last rejected.

But the search to understand the vital principle by which the body worked could still throw up some strange episodes, such as that of Franz Anton Mesmer (1734–1815), a German physician whose career became a sensation in Paris in the 1780s. Mesmer advanced the theory known as "animal magnetism," according to which health or sickness depended on an invisible fluid that was present throughout nature. When it became blocked or disturbed, sickness resulted; but this fluid could be manipulated by the skilled doctor by various physical means and by hypnotism.

Whether Mesmer sincerely believed in his system of therapeutics or whether he was a charlatan is impossible to say. But perhaps the most interesting thing about this episode is that a commission of the Paris Academy, which included Benjamin Franklin and the great chemist Lavoisier, condemned it as fraudulent—a sign of the rising authority of scientific ideas. Mesmer's name entered the English language in "mesmerism," meaning hypnotism.

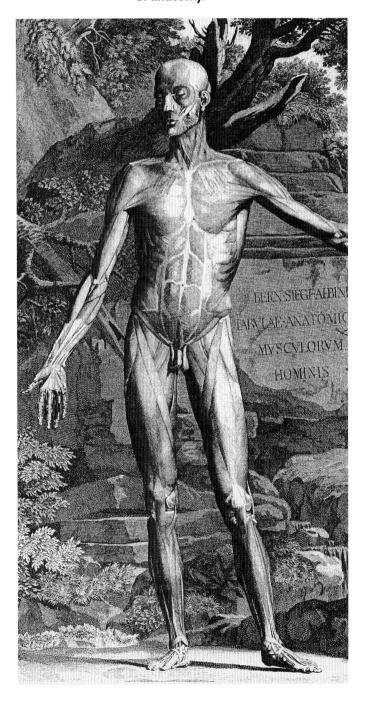

A page from Albinus's great atlas of anatomy.

Edward Jenner and Vaccination

Above: Cow horn inscribed "G C Jenner 1825." It came from a cow used by Jenner to provide cowpox-infected fluid to vaccinate people against smallpox.

Despite all its limitations, it was in the realm of medicine that eighteenth-century science achieved one of its most visible and outstanding triumphs: the conquest of smallpox by the English physician Edward Jenner. Smallpox had replaced bubonic plague as the greatest health scourge in Europe, causing thousands of deaths each year. Where it failed to kill, it left disfiguring scars. Jenner (1749–1823) was a country physician with no pretensions to scientific philosophy, and his discovery was the result of careful observation and testing.

Jenner himself did not invent the technique of inoculation: The idea that a tiny dose of an infection could build protection against a fatal dose was established early in the eighteenth century. A healthy person had matter from an old smallpox pustule scratched into their skin, with the aim of producing a mild infection. However, the technique was dangerous, for the virulence of the dose could not be controlled, and sometimes patients died.

CLOSE OBSERVATION

Jenner noticed that certain groups in the rural population, especially milkmaids, appeared to be immune to smallpox. On investigation he discovered that these people had previously contracted a mild fever known as cowpox. Working from this, Jenner tried giving healthy people a small dose of matter from a cowpox pustule. This did indeed induce cases of cowpox, but it also gave immunity to smallpox. Jenner proved this with his first human guinea pig, an eight-year-old boy named James Phipps, who showed no effects when injected with smallpox after vaccination with cowpox. Strangely enough, as Jenner noticed, the vaccination did not protect against the milder disease of cowpox itself.

In June 1798 Jenner published, at his own expense, a pamphlet in which he announced his method, and in which he coined the word <u>virus</u> to describe the agency that caused smallpox. The pamphlet electrified the medical world, and news of the discovery spread rapidly to physicians in all parts of Europe. Jenner was inundated with requests for the vaccine and questions about its workings. He had found that <u>lymph</u> taken from cowpox pustules could be dried and stored, and that it would remain effective for several months, permitting him to send it long distances. European diplomats in Turkey and India promoted the vaccine there, while

physicians in New England introduced it successfully into America. Among those who received vaccine direct from Jenner was President Thomas Jefferson.

Edward Jenner, in a stained-glass window from the house of Dr. Gunasekra in London, England.

WORLD FAME

Jenner became an international hero, and in 1804 Napoleon had a medal struck in his honor in spite of the fact that England and France were then at war. At Jenner's request Napoleon ordered the release of certain Englishmen interned in France, such was the prestige of the mild country physician. He was granted a large government pension in recognition of his achievement.

Jenner can be seen as the founder of immunology who opened up the possibility of combating other virulent diseases in a similar way. He owed his success to careful experiments and his clear powers of deduction.

He was also a keen naturalist and first described the strange habits of the cuckoo, explaining why that bird's nests had never been found, and how it destroyed the sparrow nestlings with whom it was hatched. After he had become famous, he had little rest from dealing with his vast correspondence, but he returned to his bird studies when he could, puzzling over the mysteries of migration.

EDWARD JENNER (1749–1823)
- Physician and vaccination pioneer.
- Born Berkeley, England.
- Studied medicine while apprenticed to a surgeon at Sodbury.
- 1770 Moved to London to study further, then returned to Berkeley in 1773 and worked in his own practice.
- 1775 Started studying cowpox and concluded that it would immunize against smallpox.
- 1796 Vaccinated an eight-year–old boy, James Phipps, with cowpox, then two months later inoculated him with smallpox—he did not catch the disease.
- Privately published a short explanation of his experiments in *An Inquiry into the Causes and Effects of the Variolae Vaccinae*, 1798—the scientific publications refused to publish it.
- Initially the medical profession was very hostile to the idea of vaccination, but within five years it was widely used around the world.

The New Science of the Eighteenth Century: Electricity

Petrus van Musshenbroek of Leiden.

Physical experimenters of the seventeenth century, such as William Gilbert and Otto van Guericke (see Volume Five), had discovered that a metallic sphere, when rotated or rubbed, acquired the power to attract small objects and to give off sparks. Serious interest in these effects began a few years later, when two English experimenters, Francis Hauksbee (1666–1713) and Stephen Gray (1666–1736), made better spheres out of glass that could be turned on a spindle, and they also produced sparks when touched, made objects attract each other, or gave off glowing light. They had invented the static generator.

Gray made the vital discovery that the mysterious force that they were generating could be carried hundreds of feet along some materials, such as silk, but that other materials, such as thick brass wire, would not transmit it. Thus he had discovered both <u>conductivity</u> and <u>resistance</u>. Amusing experiments were devised in which electricity was carried from the generator across a room and made to ignite vapors rising from alcohol.

Hauksbee and Gray could not explain what was happening in these experiments. They could only speculate about the existence of a "subtle fluid" that might be contained inside all matter, and that could be released under the right conditions. Considerable excitement was generated among scientists about this strange fluid.

Shortly before he died, Gray discovered that a pendulum would revolve around an electrified object. He instantly made the connection between this effect and the movement of the planets around the Sun, and he hoped to able to prove this connection by building an electric model of the solar system. "If God will spare my life a little longer," he wrote, "I hope to be able to astonish the world with a new sort of Planetarium… and with a certain theory for accounting for the motions of the Grand Planetarium of the Universe."

MYSTERIOUS FORCE

If this mysterious force were a fluid, the question arose whether it could be collected and stored like any other. In 1746 Petrus van Musshenbroek of Leiden in Holland (1692–1761) tried coating a bottle with metal paint, which he then filled with water and connected to a static generator. His aim was to analyze the liquid to see if it differed from common water. When he touched the contact to withdraw it from the bottle, he received so strong an electric shock that he vowed he would not repeat the experience "for all the kingdom of France."

This "Leiden jar," as it came to be called, was what we would call a condenser, and reports of it spread rapidly. It enabled further experiments to be carried out, and parlor games were organized in which circles of men and women sat hand in hand and became electrified. It was widely noticed that electrically charged objects repelled each other—in other words, they seemed to behave like magnets in reverse.

NATURAL POWER

In America in the 1740s Benjamin Franklin carried out many electrical experiments, and he began to suspect that lightning was actually a discharge of electricity. He set out to prove it by using a kite in a storm to draw down lightning into a Leiden jar.

The immense natural power of lightning was well known, so that when it was proved that this force was identical to that being employed in these parlor games, electricity became a subject of even greater significance. Franklin concluded that "Electrical fire was an element, diffused among and attracted by other matter, particularly by water and metals." It seemed to him that when one body built up an excess of electricity and was brought close to another body with less, a discharge occurred to equalize the two.

But how could this force be controlled or measured? In Russia Georg Richman (1711–53), a German scientist working in St. Petersburg, was trying to measure the electric force of lightning during a storm when he was killed because he had neglected to insulate himself from the ground—an event that deeply impressed his scientific colleagues.

Benjamin Franklin (Bottom) and one of his early lightning-conductors (Below). Franklin demonstrated that lightning was a discharge of electricity, and he showed how tall buildings could be protected from it by conducting the charge to the Earth.

MEASURING THE INVISIBLE

The first successful attempt to measure electricity came when the French engineer Charles Coulomb (1736–1806) devised an apparatus called a "torsion balance" that showed the force required to bring two small electrically charged bodies together. The most striking thing about Coulomb's results was that the force varied as the square of the distance between them. This was identical to Newton's law of gravity, and it raised the possibility that electricity was the force that ruled the universe, which Newton had analyzed but never actually identified or isolated. Coulomb's name was later given to the standard unit of charge, one amp per second.

ANIMAL ELECTRICITY

The final electrical experiments of the eighteenth century took place in Italy, and they were, in some ways, the most mysterious of all. In the 1780s at the University of Bologna Luigi Galvani (1737–98) tested the effects of electricity on dead animals. He found that stimulation from a low electric charge made the legs of frogs twitch. He also found that the same twitching occurred when a

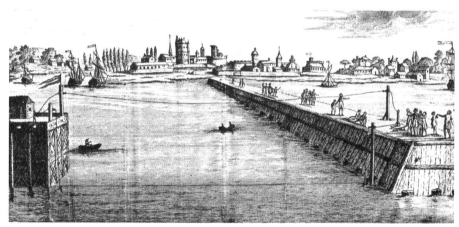

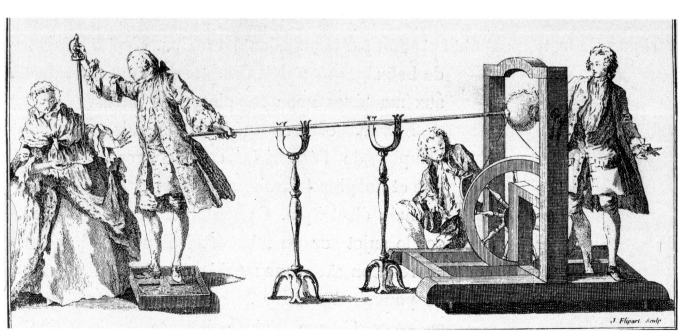

Above: Another electrical trick: An electrified boy hanging from the ceiling, and the woman on the right draws a spark from his nose.

Opposite, Above Left and Below Left: Volta (above) and Galvani (below), who disagreed about the role of electricity as the mysterious principle of life.

Opposite, Right and Bottom: Galvani's experiment that caused movement in the muscles of dead frogs.

ALESSANDRO VOLTA (1745–1827)
- Physicist and prolific inventor.
- Born in Como, Italy.
- 1775 Appointed professor of physics at Como University, then in 1778 at Pavia.
- He devised the electrophorus—an early induction machine, 1775; invented an electric pistol using "inflammable air" (hydrogen) in 1777; the condenser, 1778; the candle flame collector of atmospheric electricity, 1787; the "voltaic pile" (the electrochemical battery)—the first supply of continuous electricity, in 1800.
- 1795 Became rector of Pavia University, but four years later he was fired for political reasons, although later reinstated by the French.
- The volt, an SI unit of electrical potential difference, is named after him.

copper probe was touched on the spinal cord, and the copper then made contact with iron, especially damp iron.

Galvani though he had discovered "animal electricity," and that he had confirmed what physiologists had suspected since the time of Descartes, namely, that the nerves were filled with a subtle fluid, directed from the brain, that caused muscular action, and that this fluid was none other than electricity. The word "Galvanism" was immediately applied to this force.

The principle appeared to be supported by naturalists, who had now realized that the shocks that came from eels and rays were also identical to electrical effects.

However, at the nearby University of Pavia Galvani's friend Alessandro Volta (1745–1827) repeated these experiments, but came to very different conclusions. Volta believed that the twitching of the muscles was actually caused by the contact across a damp medium between the two different metals of the probe and the plate on which the frogs were mounted; the electricity did not come from the animal at all. He verified that electrical effects would result from two metals in contact in a damp environment—for example, if the tongue is touched with a bimetallic strip, a tingling sensation is felt.

FIRST BATTERY

The culmination of Volta's proofs came when he built his "Voltaic pile." A series of metal disks, alternately zinc and copper, were separated by paper soaked in brine or weak acid, connected by a metal strip. This pile produced a small, steady electrical current that could be strengthened by adding more piles in a series. This was the first battery or electric cell, and it offered a source of electricity apparently unconnected with the static machines.

This electricity was easier to produce and handle than static electricity, but it produced yet another surprise: When the Voltaic pile was immersed in water, bubbles of gas were seen rising from the plates. When these gases were collected and analyzed, they proved to be oxygen and hydrogen, while the water gradually disappeared. The only possible conclusion was that some *chemical* reaction was occurring that was also capable of producing electricity. The implications of all this were deeply puzzling. Electricity appeared to be a powerful natural force locked up inside matter that could be released by a variety of means, both physical and chemical. A further series of experiments by a new generation of scientists would be required before these effects were brought together into a clearer understanding of electricity, but there was no doubt that a major new physical force had been identified, one that might be still more important than magnetism, air pressure, or even gravity itself.

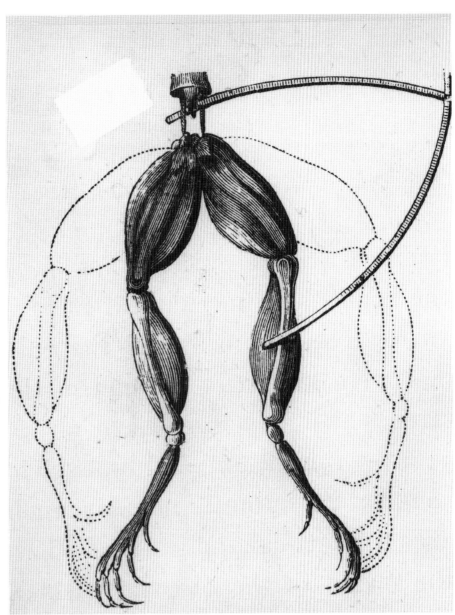

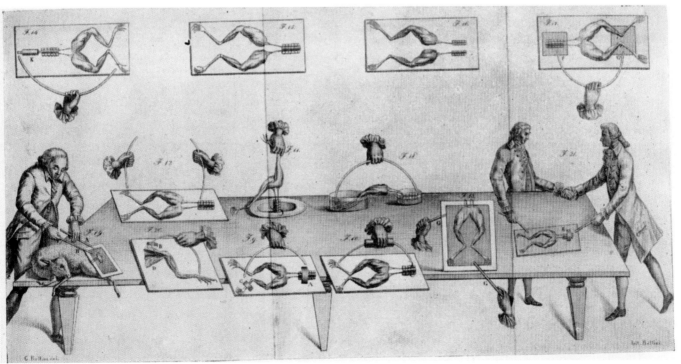

The Steam Engine:
The First Machine

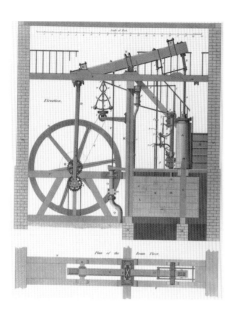

JAMES WATT (1736–1819)
- Engineer and inventor.
- Born in Greenock, Scotland.
- Educated at Greenock grammar school.
- 1754 Went to Glasgow to learn how to make mathematical instruments.
- Started studying steam as a motive force in about 1759.
- 1757–63 Worked as mathematical instrument maker at Glasgow University.
- 1767 Helped survey the land for the Forth and Clyde Canal, then worked on the deepening of the Forth and Clyde rivers and on improving the local harbors.
- 1763-4 While repairing a model of Newcomen's steam engine in his workshop, devised the separate condenser—making it three times more effective.
- 1774 Entered into partnership with Matthew Boulton to manufacture the new engine at Soho Engineering Works, Birmingham.
- 1769 Patented his condenser and went on to patent other developments: sun-and-planet motion, 1781; the expansion principle; the double acting engine; the parallel motion; a smokeless furnace; the governor, 1785.
- 1783 Specified and named the original unit of power—the horsepower.
- 1784 Patented a steam locomotive but did not pursue the idea.
- The metric unit of power, the watt, is named for him

The eighteenth century was an age of theory and of experiment, and it was an age when all educated people were becoming aware of the new scientific ideas. But as yet, science had little practical impact on society: Technological change had not yet revolutionized people's lives or their environment. The reason for this was that no new power sources had been discovered. In fact, the very concept of a "power source" would have been puzzling. Anything that had to be moved was moved by muscle power, human or animal, and the only exception was on water, where wind or current could be exploited. The only technical device that employed an unusually large force was the cannon, and it had no application in ordinary life, its power being explosive and destructive.

Yet at the same time as scientists were theorizing about phlogiston (see above, page 27), about animal electricity, or spontaneous generation, an almost unnoticed revolution was under way in engineering that would completely redefine the role that science plays in society. The steam engine was invented in the context of practical mining, and its pioneers had no contact with the scholars of the scientific academies.

ENGINEERING BREAKTHROUGH

One of the perpetual problems of mining was to remove water from the mine shafts; small pumps had to be worked continually by hand. It was Thomas Savery (1650–1715), an English merchant and engineer, who conceived the idea of a mechanical pump, which he first described as "an engine to raise water by the agency of fire." The principle involved was the knowledge that when steam is condensed, a partial vacuum results. The resulting machine was a forerunner of the steam engine.

Savery's invention was more of an apparatus than a machine, for it had no moving parts other than valves, which were operated by hand. Steam from a boiler filled a pipe, which was then cooled externally by cold water. The resulting vacuum would draw up water from a mine, but only to a maximum height of around 30 feet because of atmospheric pressure. The cycle could be repeated four or five times per minute. Savery received a patent for this design as early as 1698; but as far as we know, only four devices of this type were built in England, and there are no reports about how effective they were.

PRACTICAL APPLICATION

It was from this apparatus that a more complex type of machine was developed by Thomas Newcomen (1663–1729), another merchant and practical engineer. The first "Newcomen engine" or "beam engine" was built at a mine in Warwickshire in 1712. It differed from Savery's in that the vacuum was made to work a piston connected to a large horizontal beam, the other end of which raised and lowered the pumping rod. The hand-operated valves were eliminated, and the cycle now worked up to 12 times per minute. While steam forced the piston up, the partial vacuum formed after cooling brought it down again. Several hundred of these engines were built at mines in England and in continental Europe, and they worked for half a century and more. Nothing like them had ever been seen before, and their power must have seemed immense compared with that of hand-operated pumps.

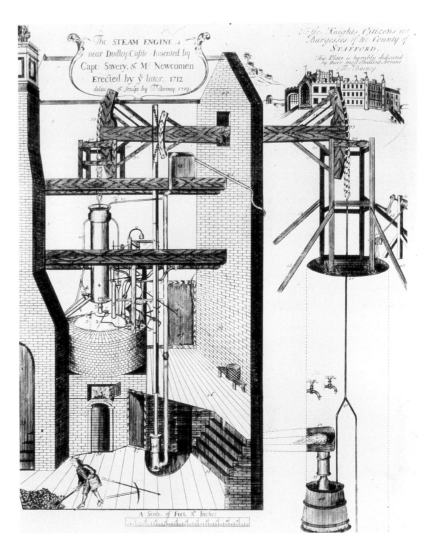

In modern engineering terms the Newcomen engine was desperately inefficient because it was alternately heated and cooled, wasting most of its huge input of fuel. Critics complained that it required the product of an iron mine to build one of the machines and the product of a coal mine to drive it. There was also no appreciation that this source of power could be put to other uses than raising and lowering a pumping piston.

Above: A Newcomen beam engine, improved from Savery's model to work a huge lever, which raised and lowered a pumping piston.

Opposite: A James Watt steam engine of the 1760s, which really fuelled the Industrial Revolution. The beam now turns a great flywheel.

GREATER EFFICIENCY

This situation was transformed by the Scottish engineer James Watt in the 1760s. Watt (1736–1819) achieved four great innovations in his new design. First, he introduced a second chamber to condense the steam separate from the piston chamber, so that the latter could always remain hot. This saved three-quarters of the fuel used by a Newcomen engine. Second, he used the pressure of the steam itself to force the piston down as well as up, creating a much stronger double action. Third, and somewhat later, he transformed the up-and-down action of the beam into rotary action by gearing it via cranks to a flywheel. The fourth innovation was the ingenious "governor," the valve that was automatically closed when the machine reached a certain speed. In some ways the governor was the most remarkable feature of all, because it worked by feedback from the machine itself to make the whole system *self-regulating*.

WATT.

WIDER APPLICATION

The result of Watt's brilliant design was a vastly improved power source, and one that could be put to many more uses than pumping up water. Rotary action steam engines could be used to turn lathes, saws, mills, textile machines, wheels, ships' propellers, and a dozen other applications. Steam engines became so prevalent in all fields of engineering by the end of the eighteenth century that it became necessary to grade their power output. The unit chosen was the measure called the "horsepower," and it was defined as the ability to raise 33,000 pounds weight one foot in one minute against the force of gravity. In the United States it is now defined as equal to 746 watts.

Above: Newcomen atmospheric engine of 1730—one of a series of French technical drawings of engines erected in waterworks on the Thames River. The engine gained its pressure by using the vacuum created by steam and fire.

Opposite: James Watt, who transformed the Newcomen beam engine into a versatile source of power with many applications.

NEW INSIGHTS

There were three outstanding results from the invention of the steam engine. First, the living and working environments of millions of people were changed as their lives became geared to what machines could perform. Second, technology acquired a new momentum of improvement, since there was a constant search for better and better machines capable of more and more applications. The limitation of muscle power was a thing of the past: The very idea of the machine led to more machines. This process was fed by growing scientific and engineering skills and, ultimately, by the desire for commercial advantage. Third, there was the intellectual impact on the understanding of physical forces.

Scientists began to analyze the way in which the steam engine actually worked—how, to use Savery's quaint words, "water could be raised by the agency of fire." The answers at which they arrived gave them nothing less than a new understanding of physics. They led to a whole new scientific language that took shape in the nineteenth century, and that we now call classical physics. The machine age had arrived, it seemed, almost by chance, but it would lead to vital new insights in the realm of pure science.

Theories of the Earth

Just as eighteenth-century scientists produced new theories about the working of the human body or the relationship between the many species of animals and plants, so they turned their attention to the Earth itself. Traditional thought on this subject was dominated by the biblical narrative, which taught quite simply that the Earth and everything on it had been created at one moment in time in the remote past. By tracing back the chronological sequence of names and events given in the Bible, scholars believed they had been able to fix the date of the creation at 4004 B.C. The story of earth science in the eighteenth century is the story of the way this doctrine of instantaneous creation was questioned, modified, or rejected. It was gradually replaced by a view of a *changing* Earth, whose features and life forms had been shaped over a much longer period of time than that implicit in the Bible.

CREATURES IN THE ROCK

One of the key pieces of evidence was the presence of fossils inside rocks. That fossils resembled known animals—especially sea creatures—had long been recognized, but it was believed that this was merely an accident of nature. It was Niels Stensen in the 1660s (see Volume 5) who argued that they did not merely resemble living creatures, but were actually their petrified remains. But how was it that some of these fossils were of animals no longer living on the Earth? Could some species have died out entirely? Strangely, the Bible came to the aid of naturalists here with the story of the flood. The unrecognized life forms must surely be the remains of those that were destroyed in that cataclysm. This explanation was advanced by the Swiss naturalist Johann Scheuchzer (1672–1733), who gathered a large collection of fossils from his journeys in the Alps. He was convinced that the presence of fossilized sea creatures high in the mountains showed they had been deposited there by the biblical flood.

The idea of harmonizing science and the Bible was taken

A junction between strata of rocks formed at quite different epochs, drawn by Hutton. He reasoned that the lower level could not have been laid down vertically, but must have been heaved up by violent movement in the Earth.

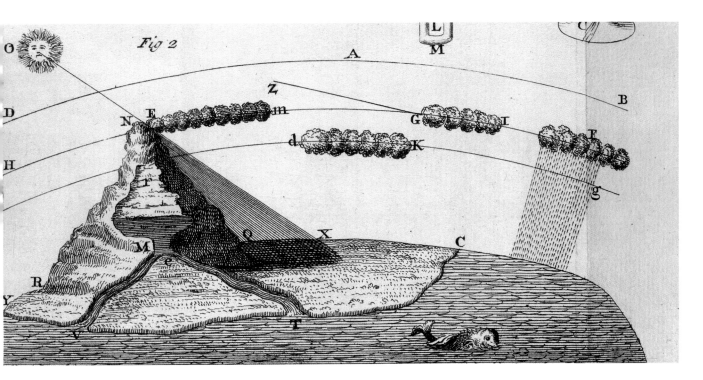

Fig 2

up by, among others, William Whiston (1667–1752), who was Newton's successor as professor of mathematics at Cambridge University in England. Whiston proposed that the flood had been caused by a comet striking the Earth, and that other miraculous events in the Bible could be given scientific explanations. This "demythologizing" of the Bible angered orthodox thinkers, and Whiston was driven from the university.

The water cycle, from a treatise of 1729. Eighteenth-century scientists correctly understood that water evaporates and recondenses in an endless cycle; earlier it had been believed that rivers arose from vast underground seas, or that the oceans themselves fed the rivers through cave systems.

NEW THEORIES

It was Buffon who took the decisive step of rejecting the biblical chronology outright. One of Buffon's central theories (see page 20) was that the Earth was formed of solar material torn off when a comet struck the Sun. Therefore the Earth had been undergoing a cooling process ever since. This was relevant to the problem of fossils because it was evident, Buffon claimed, that certain creatures had once lived that were adapted to much higher temperatures, but that had died out as the Earth cooled.

While Buffon was writing in the 1750s and '60s, some large fossilized land animals—such as mammoths in Siberia—had been discovered. Buffon contended that the northern regions of the Earth had once been as warm as the tropics, and that as the Earth cooled, these animals had migrated south and developed into elephants. This entire process of cooling and its complex results must have occupied tens of thousands of years, and Buffon stated explicitly that the age of the Earth was of a very different order from that suggested by the Bible.

FIRE VERSUS WATER

How had all the different rocks and minerals and features such as mountains, oceans, and deserts arisen? By the second half of the eighteenth century there were two main schools of geological thought, one asserting the primacy of water, the other the primacy of fire.

The first school came to be known as Neptunism (after the classical Roman god of the sea), and its most influential figure was the German geologist Abraham Werner (1749–1817). Werner thought that ocean had once covered the entire Earth, and all the rocks and minerals that now exist had been formed as this ocean evaporated and retreated. The older rocks, such as granite, had been crystallized from chemicals in the ocean and emerged first. The younger, sedimentary rocks were the products of erosion and were found superimposed on the older rocks. Volcanic rocks were the latest of all. The strength of Neptunism was that it recognized the principle of stratification; its weakness was its uncertainty about the time scale involved. Some advocates of Neptunism still identified the ocean-covered Earth with the period of the biblical flood and welcomed the theory as confirming the Bible.

The opposing theory was known as Vulcanism (named after the Roman god of fire) because it saw volcanic disturbance as the primary force that had formed the Earth's crust. Its best-known advocate was the Scottish philosopher James Hutton (1726–97), whose book *Theory of the Earth*, 1795, was one of the most influential books in early geology.

In Hutton's theory subterranean heat threw up land masses that were continually worn down by erosion. This fiery activity was like a chemical laboratory on a huge scale, producing all the rocks and minerals that were to be found on Earth. Fire, uplift, and erosion formed a never-ending cycle. Perhaps the most striking aspect of Hutton's theory was his insistence that the processes that had formed the Earth were still at work now: Land was being formed and eroded over vast tracts of time, and this proved that the age of the Earth must be inconceivably great. Vulcanism was generally nearer to the truth, as established by later geologists, than Neptunism.

RAIN FLOW

A related problem was the hydrological cycle—where did rivers arise? Why did rivers and rainwater not cause the sea to rise and flood the land? There must be a cycle in which the amount of water remained constant but took different forms. The older theory supposed a subterranean world of a huge network of caves and pools through which seawater flowed in hidden channels back to the sources of all the world's rivers, somehow losing its salinity in the process.

Edmond Halley (1656–1742), a colleague of Newton and a comet analyst, took a much more scientific approach to the problem. He calculated the discharge of the main European rivers into the Mediterranean and compared it to the amount of water lost by evaporation each day. He concluded that the evaporated water condensed and fell as rain that swelled the rivers and returned to the sea. Other scientists repeated this approach on a smaller scale—for example, with one river basin—and found it to be correct.

Johann Scheuchzer, the Swiss naturalist, collecting fossils in the Alps. He argued that they were the remains of life forms that no longer could be found on Earth—many of them having been killed by the biblical flood.

Scientific Expeditions

The empirical ideal of knowledge in the eighteenth century created a new form of scientific investigation: the organized expedition. For centuries Europeans had been exploring the world for reasons of trade or conquest, but now there was a new motive. Expeditions set out with the aim of enlarging geographical knowledge, collecting specimens of unknown plants and animals, observing new physical features on the Earth's surface, or studying exotic peoples.

GOING HUNTING

The scientists of the time realized that no theory, whether it concerned chemistry or physics, biology or earth science, could be securely based until all the data were examined. They realized, too, that large areas of the globe had never been properly explored, and this was a situation that was deeply unsatisfactory. In the face of great obstacles—extremes of climate, hostile peoples, lack of maps—they set out to enrich the body of knowledge available to science.

The case of South America was typical. The continent had been settled by the Spanish and the Portuguese for almost 200 years, but their interest in it was purely economic—to extract precious metals or hardwoods, or to farm sugar. When the French expedition arrived in Peru in 1736 to measure the arc of the meridian (see above, page 11), many of the scientists were fascinated by the strange environment in which they found themselves. When their task was completed, the expedition leader, Charles de la Condamine, set out on a pioneering voyage down the Amazon River.

He noted the richness of the flora and fauna, measured the immense volume of water in the Amazon basin, studied the native tribes, and experimented with their poisoned arrows. La Condamine was the first person to alert European naturalists to the potential of South America as a natural laboratory, and he first described to Europe the strange quality of rubber, which he found secreted from various plants.

An offering before Captain Cook. Natives of Hawaii slaughtering swine during Cook's third and last Pacific voyage (1776–79). The main aim of this voyage was to find a northern passage between the Pacific and Atlantic oceans. Cook would die in Hawaii after his ship *Resolution* returned to the island after storm damage in 1779.

One of La Condamine's colleagues, Pierre Bouger, noticed that the swing of a pendulum and of the needle of a magnet were both affected by the mass of the Andes Mountains, and the name "Bouger anomaly" was later given to this phenomenon.

Bouger effects are localized and not to be confused with magnetic variation, that is, the angular difference between true north and the north magnetic pole. Magnetic variation had already been carefully measured by the English scientist Edmond Halley, who undertook a long Atlantic voyage in the years 1698–1700 for that purpose and later published a chart that became essential equipment for every navigator.

Above: Volcanic material from Mount Vesuvius, Italy, 1779. This illustration (and that at left) is one of five from the 1779 supplement to William Hamilton's study of Italian volcanoes. The most precise study of volcanoes up to that time, Hamilton's identified that volcanoes are created by natural forces.

Opposite: The eruption of Mount Vesuvius on the morning of Monday, August 9, 1779.

MAPPING THE STARS

This was not Halley's first scientific voyage, for in 1676, when he was only 20 years of age, he had sailed to the island of St. Helena in the south Atlantic in order to enlarge the map of the heavens. When he returned, he had charted the positions of several hundred new stars invisible from northern Europe.

Strangely perhaps, this process was not carried further until 1750, when the French astronomer Nicolas de Lacaille spent a year surveying the southern skies from the Cape of Good Hope. Lacaille not only charted new stars, but formed them into 13 new constellations that have appeared on celestial maps and globes ever since. Working in the spirit of eighteenth-century science, Lacaille turned his back on the traditional animal figures and formed his constellations as technical subjects—the clock, the compass, the telescope, the pump, and so on.

PACIFIC HIGHWAY

The region of the world that most attracted scientific expeditions was the Pacific Ocean, the largest unexplored area of the globe. There were two motives at work here: first, to complete the map of the ocean itself, and second, to study its climate, wildlife, and peoples. There was a particular problem with the Pacific in that European geographers had believed for centuries that a great southern continent must exist there to counterbalance the land masses in the northern part of the globe.

Englishman Captain James Cook (1728–1779) undertook three great voyages between 1768 and 1779 on which he explored from the waters of the Antarctic to the Bering Strait and discovered many islands groups, including Hawaii. Yet so strong was the belief in the "great southern continent" that he was roundly criticized by armchair geographers in Europe for failing to find it.

Cook was accompanied on his first voyage by Joseph Banks, a wealthy naturalist who collected hundreds of specimens of fish, insects, plants, and animals. Banks later became president of the Royal Society in England and sponsored many more overseas scientific expeditions.

FAR-FLUNG PEOPLES

What Cook did find was the puzzle of a mass of far-flung islands inhabited by people with a common culture. How had these people become spread across such vast distances?

Some scientists explained the phenomenon of the Polynesian peoples by theorizing that the Pacific islands must be the remains of a continent that had sunk beneath the ocean centuries ago.

The people themselves presented an intellectual problem to Europeans: They lived apparently without laws, without property, without kings or governments, and they were free of the great social evils of crime, war, and misery that plagued the European nations. They appeared to be "noble savages," but how could this be reconciled with the Enlightenment belief that civilization was based on reason, law, science, and knowledge?

The Pacific region was sometimes spoken of as a new Eden rich in unknown life forms, including a race of people who had not lost their innocence. Of all the new species observed in the Pacific, it seemed that the human inhabitants provoked most thought, and the study of anthropology really began when Europeans were confronted by these mysterious people.

CROSSING CONTINENTS

Expeditions overland were, in many ways, more difficult than

Opposite: *Endeavour,* around 1770, laid on the shoreline of New Holland (Australia) for hull repairs.

Below: Map showing Captain Cooke's voyages, dated around 1784. His first trip, in the *Endeavour,* was to observe the transit of Venus and seek out the fifth continent—Australia; the second and third (in *Adventure* and *Resolution* respectively) were to find a northern passage.

Bottom: A chart of Tahiti around 1769, showing the route of Captain Cook's first voyage to the south Pacific.

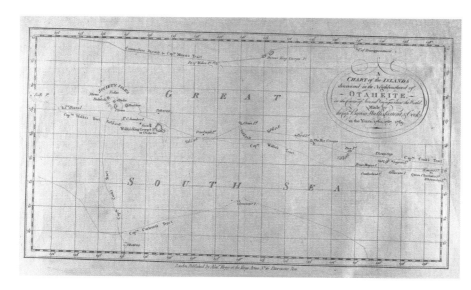

by sea. The African interior would resist exploration for another century. However, one great region that was accessible was Asia, and since the sixteenth century the czars of Russia had been gradually extending their control eastward.

In 1728 Vitus Bering (1681–1741), a Dane in the Russian navy, commanded an expedition to establish whether Asia was, or was not, attached to America in the extreme northeast, and he first sailed through the narrow strait that now bears his name.

Forty years later the German scientist Peter Simon Pallas (1741–1811) was commissioned to explore Asian Russia from the Crimea to eastern Siberia. Pallas discovered the remains of extinct mammoths frozen in the Siberian ice; he was especially interested in the relation between animals and their environment, showing how climate and food affected their distribution, and he advanced Vulcanist theories about mountain building.

On all these expeditions scientists deliberately set out to extend their horizons of knowledge. If new theories sprang from them, so much the better, but it was the collection of the data that became a scientific ideal in itself. The Earth came to be seen as a vast natural laboratory that must be explored and understood. An essential part of the work was that their reports should be published for others to study. The lesson of these expeditions was that the Earth contained many different *environments* in which climate, animals, plants, soils, rivers, and people all interacted in different ways. Only when this principle became generally understood did geography itself develop during the nineteenth century as the science of the environment.

JAMES COOK (1728–79)
- Navigator and explorer.
- Born in Marton, England.
- Apprenticed to a ship owner in Whitby. Learned shipcraft working as a seaman along the North Sea coast and in the Baltic.
- 1755 Joined the Royal Navy, getting his master's license in 1759.
- Sent to survey the coast and seas around Newfoundland and the St. Lawrence River.
- 1768–71 Commissioned by the Royal Society for an expedition to the Pacific on board the *Endeavour* to observe and record the transit of Venus across the Sun.
- Came back via New Zealand, Australia (which he claimed for Great Britain), New Guinea, Java, and the Cape of Good Hope.
- 1772–75 Promoted to commander and given the *Resolution* and *Adventure* to investigate the extent of the Antarctic ice. During this trip also visited Tahiti and the New Hebrides, and discovered New Caledonia and other islands.
- 1776–79 Sent to discover the northwest passage around America from the Pacific. Sailed via the Cape of Good Hope, Tasmania, New Zealand, the Pacific, and Sandwich Islands to arrive at the west coast of North America. Surveyed as far as the Bering Strait.
- 1779 Returned home via Hawaii, where he stayed for a month before being killed by islanders.

Astronomy: Looking beyond the Solar System

Thomas Wright's vision of multiple galaxies. Wright, an English instrument-maker and astronomer, believed that star groups such as our own might form discrete systems, and that many such star worlds might exist throughout the universe.

Classical astronomy had always concentrated on the solar system, analyzing the paths of the Sun, Moon, and planets. The stars had been catalogued and marked on the chart of the heavens; but they had always been seen as mere points of light, and it had been assumed that humans could know nothing about them. Traditionally, people thought they were set in a single "starry sphere" all at the same distance from the Earth. This realm of the universe was so remote and unknowable that it was regarded in religious terms as the abode of God and his angels.

By the end of the seventeenth century the Copernican revolution and the advent of powerful telescopes had altered these ideas. It was now accepted that the Sun was a star, and the other stars were suns. This insight raised profound questions about the scale and structure of the universe. The eighteenth century saw the beginning of stellar astronomy, although at first of a rather speculative kind.

IMMEASURABLE UNIVERSE

The first aim was to arrive at a more accurate idea of the scale of the universe. Many astronomers, from the time of Newton and Huygens, had suggested ways in which improved figures might be obtained for the distances between the Earth and the Sun and the Earth and the stars. These methods usually involved calculations based on <u>parallax</u> and on relative brightness. None of these figures ever came close to accuracy, but they all exploded the old view of the size of the universe that had prevailed since Ptolemy, and scientists became keenly aware that the universe was immeasurably vast.

It was Edmond Halley, who made important contributions to so many fields of science, who first questioned the doctrine that the stars were fixed and immovable. In 1718 he published observations that suggested that the stars had moved over time, making the positions given in classical star catalogues from Ptolemy to Hevelius slightly inaccurate. No general pattern in these movements was detectable, and Halley concluded that the stars were moving freely, perhaps randomly, in space.

The implication of this observation and discovery was that the stars were *scattered* throughout space at varying distances from the Earth, so the old idea of the celestial sphere as a shell bearing all the stars was a myth. The question then arose whether the universe had some large-scale structure, and, if so, how observation by Earthbound astronomers could uncover it and begin to understand the cosmos and ultimately prove its existence scientifically.

CHANGING POSITION

That the Sun was not in any sense the center of the universe was now accepted by all scientists, and it seemed logical to believe that the Sun itself was also moving through space, carrying the Earth and the entire solar system with it. All astronomers of this time made one important assumption—that all the stars were of approximately the same size and luminosity, and that the brightness or faintness of a star indicated fairly directly how far from the Earth it lay.

For the next 50 years astronomers struggled with the near-impossible task of measuring changes in stellar positions. Before the work of William Herschel (see below, page 56) there were two major breakthroughs.

Newtonian reflecting telescope (see Volume 5, page 25) on an altazimuth stand. The telescope was made by the famous astronomer William Herschel (see page 56) for his friend Sir William Watson.

Astronomy in fashion: studying and marveling at the beauty and order of the heavens were a popular pursuit among all educated people.

In 1729 Englishman James Bradley (1693–1762), who succeeded Halley as Astronomer Royal at the Greenwich Observatory, announced that the positions of all stars varied by infinitesimal amounts from their predicted values *on an annual cycle*. Some values were in advance, some behind, but in both cases they corrected themselves before creeping in again. Bradley argued that the only logical explanation was that the velocity of the Earth in its orbit around the Sun was being added to or subtracted from the finite velocity of the light incoming from the stars.

This was the "aberration of light," and it was important for two reasons. First, it gave the first physical proof of the Copernican theory that the Earth is moving; second, given the immense velocity of light, it indicated just how vast the distances to the stars must be.

TRAVELING THROUGH SPACE

Another English astronomer, John Michell (1724–93), drew attention to double stars—binaries as they are now called—and argued from them that gravity operated far outside the solar system. In 1760 the German scientist Tobias Mayer (1723–1762) announced his discovery of a very important principle. If the Sun, and the Earth with it, were moving through space, then we should be able to detect this as stars appearing to open before us and close behind us. This effect is exactly what happens if we walk through a wood: Trees in front that had appeared close together seem to widen out before us; but, if we look behind, the wood has assumed its dense appearance again.

Mayer argued that there should exist a "solar apex" in the direction of travel and a "solar antapex" behind. This was soon confirmed, and it was established that the Sun and the solar system are moving in the direction of the constellation Hercules. The discoveries of Bradley and Mayer were a considerable triumph of astronomical thought, overturning the doctrine of the fixed stars that had been universally believed just a few decades before.

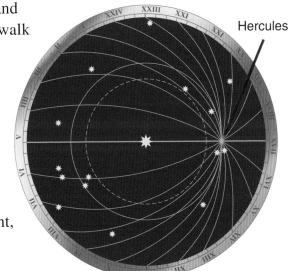

DIVINE INFLUENCE

This new interest in stellar astronomy stimulated many theories about the size and structure of the universe. Among them was the highly imaginative scheme of Thomas Wright (1711–86), an English navigator and surveyor who combined scientific vision with religious belief. Wright took the Newtonian principle of gravity and the newly discovered movement of the stars, and wove them together into a new cosmic structure.

He proposed that the universe contained a "divine center" that acted as a gravitational center around which all the stars, including our own Sun, move in orbit. This explained why the universe did not collapse into a single body under the force of gravity, a problem that had worried many post-Newtonian thinkers.

The stars, he thought, were set into a series of concentric shells, all revolving around the divine center, thus accounting for the stellar motions seen by Halley and for the aberration of light. One might see Wright's vision as a huge stellar equivalent to the solar system, magnified on a vast scale, but still working under the influence of gravity, and still with a clear center and a spherical structure.

Wright made no contribution to precise astronomy in the way that Bradley or Mayer did, but his book *An Original Theory or New Hypothesis of the Universe,* published in 1750, influenced many other thinkers. It shows that cosmological thought was moving in new directions. Wright was trying to come to terms with a universe whose scale was vast, and possibly even infinite. But he felt still that it must possess a clear and rational structure and be under divine control. Wright, like William Derham (see above, page 12), believed that God's power now exerted itself over a far vaster universe than just this solar system, and that many worlds existed scattered throughout space.

Above: Herschel's drawing of the motion of the Sun and solar system toward the constellation Hercules, as suggested by Tobias Mayer.

Thomas Wright's business-card, advertising the quadrants and other instruments that he made ("for sea and land according to the best and latest improvements")— and emphasizing that he was mathematical instrument maker to his royal highness George, prince of Wales.

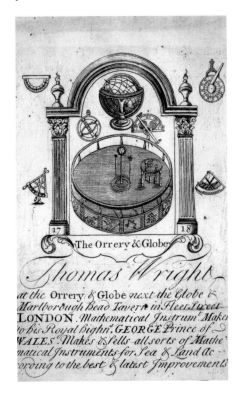

New Planets and New Stars:
William Herschel

Above: Observing a meteor from the terrace of Windsor Castle, England, August 18, 1783.

Opposite: Herschel cast his own 48-inch-diameter mirror for this giant 40-foot-long reflecting telescope.

The universe of the eighteenth-century astronomer did not consist only of stars and planets. Comet hunting became a compulsive activity because these erratic, wandering bodies were now understood to be part of the solar system and to conform to Newton's orbital laws.

It was a dedicated French comet hunter, Charles Messier (1730–1817), who turned his attention to another group of objects in the night sky. They were certain unusual stars that ancient astronomers had described as "cloudy" or "milky"—the nebulas.

Messier wanted to chart them in order to avoid any possible confusion with comets, and by 1784 he had catalogued 101 of them. They are still known according to their M numbers: The one in Orion is M42, in Andromeda M31, and so on. Messier had no idea what the nebulas were, and his interest in them was limited (although he produced some intriguing first drawings of their shapes). But his studies provided the starting point for a fresh field of astronomy, one that, much later, in the twentieth century, would become central to our understanding of the universe.

HERSCHEL'S CONTRIBUTION

Messier's work on nebulas was enormously extended by the greatest observational astronomer of the eighteenth century, William Herschel (1738–1822). Born in Germany and originally a musician, he moved to England when he was only 19. Herschel trained himself in both astronomy and in making his own telescopes, to such effect that by 1780 his instruments were the best in Europe.

In March 1781, using a seven-foot reflecting telescope, Herschel identified a body in the sky that he knew was not a star, although it might have been a faint comet. Over the following months, however, observations convinced Herschel that it was a planet, the first new planet to be added to the solar system since the dawn of astronomy 5,000 years before. Patriotically, Herschel wanted to name it *Georgium Sidus* ("George's Star") after the king of England, but other astronomers found this completely out of keeping with the classical names of the other planets. Some referred to it as "Herschelium," but eventually, the name Uranus was agreed on, in Greek mythology the father of Saturn and grandfather of Jupiter. Herschel became famous, was awarded a royal pension, and was able to build even larger telescopes and turn his attention his long-term goal in astronomy.

AMBITIOUS UNDERTAKING

Herschel's goal was nothing less than an attempt to map the universe in three dimensions. All conventional star charts to that date had treated the heavens as a two-dimensional plane in which objects are separated only by the angular distances between them. Herschel wanted to survey the radial distances of the stars from the Earth and their distribution, and so offer a picture of

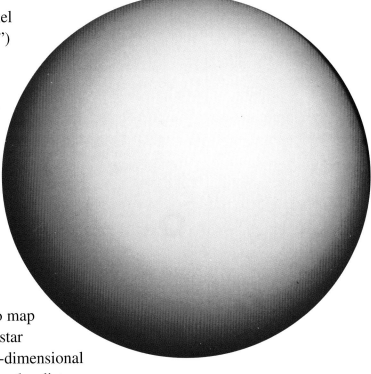

Above: Uranus, discovered by Herschel in 1781. This image was taken by Voyager 2 in 1986.

WILLIAM HERSCHEL (1738–1822)
- Astronomer.
- Born Hanover, Germany.
- Trained as a musician and joined the Hanoverian Guards band as an oboist.
- 1755 Moved to England to work as a musician.
- 1766 Settled in Bath and started an interest in astronomy.
- Taught himself to cast mirrors and built his own reflective telescope.
- 1781 Discovered the planet Uranus which he named *Georgium Sidus* in honor of King George III.
- 1782 Appointed Astronomer Royal.
- 1787 Working with his sister Caroline, he built larger telescopes and discovered two satellites of Uranus and two satellites of Saturn, 1789.
- During his studies of the stellar universe he drew up the first catalogue of double stars in 1782; proved that they orbit around each other, 1802.
- 1783 Recorded the Sun's motion through space.
- 1784 Published a paper, "On the Construction of the Heavens," revealing the Milky Way as an irregular collection of stars.
- Began a systematic search of the skies for nebulas and star clusters—discovered 2,500—published the data in three catalogues, 1786, 1789, and 1802.
- Made distinctions between the different types of nebulas.

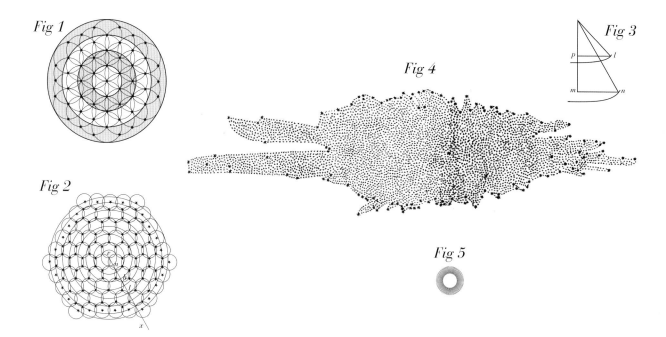

Fig 1

Fig 2

Fig 3

Fig 4

Fig 5

Above: Herschel's proposed model of our Milky Way galaxy, seen as a disk seen from the top (Figure 1 and Figure 2) and edge on (Figure 4). From the *Transactions of the Royal Society* of 1811.

Below: William Herschel, pioneer of observational astronomy beyond the solar system.

the universe in depth to determine whether there was a large-scale structure such as that envisaged by Thomas Wright.

Herschel had at that time no reliable data about stellar distances, but he made the basic assumption that all stars are more or less equal in brightness, and that the luminosity that we perceive is therefore a result of their distance from the Earth. Herschel then proceeded to count stars—thousands and thousands of stars—dividing the night sky into small set areas that he called "gages." He continued with this work throughout the 1780s, and he reached the conclusion that the stars were grouped together in a vast and rather irregular disk formation. The concentration of stars that we see as the Milky Way is our edge-on view of the disk, while when we turn away from the disk's center, the stars noticeably thin out. In this he was correct, although he supposed that all the objects that he saw in the heavens formed part of a single system.

DEEPER INTO SPACE

In 1789 Herschel began using the largest telescope of his own construction, a 40-foot reflector. It was so powerful that he began to have doubts about his model of the universe, for wherever he turned his new instrument, more stars were brought into view. He began to see that his earlier survey had not penetrated to the limits of the universe, as he had hoped. What he did not conceive was that some of the objects that he saw might be part of further and more distant systems.

Among these objects were the nebulas, to which he paid special attention. He could distinguish that some nebulas were composed of individual stars, while others were truly nebulous or cloudlike. Because they appeared in various intermediate stages between stars and clouds, Herschel suggested that what was revealed here was the life cycle of the stars. As in the nebular theory advanced by Laplace (see above, page 11), the clouds of gas were thought to be in the process of condensing into stars.

In this Herschel was mistaken, for most of the nebulous masses that he was seeing were indeed star groups too distant to be resolved. They were, in fact, galaxies beyond and separate from our own. Nevertheless, Herschel had proposed the idea that change on a cosmic scale was at work in the farthest corners of the universe. This was a profound shift from the static, mechanistic view of the universe that had prevailed at the beginning of the eighteenth century. In some ways it can be seen as the counterpart to the new Buffon-style view of nature here on Earth as in a process of gradual change. Despite the limitations of his time, William Herschel's work stands at the beginning of modern observational cosmology.

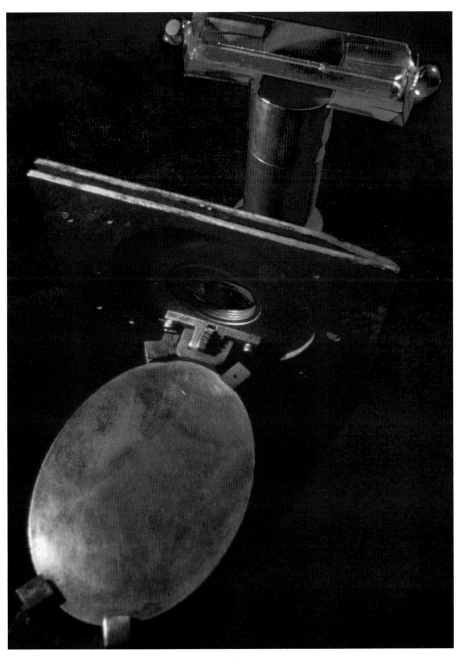

Left: Herschel's prism and mirror of around 1799. He used them in his—and the world's—first experiments into the spectrum beyond the visible region. The prism was used to observe objects under a microscope in different light. Herschel found that maximum heating and lighting effects are at different ends of the spectrum, and that the greatest heating effect is just beyond red but has no effect beyond violet.

The *Encyclopédie:* Science and Philosophy

In 1751 there was published in Paris, France, the first installment of a new encyclopedia that would eventually fill 28 volumes, and that enshrined the spirit of eighteenth-century rationalism. This huge project covered the realms of history, literature, science, art, and philosophy. It did not simply describe its subjects in a neutral or colorless way, but set out to evaluate the ideas of the past. Its guiding principle was that reason and knowledge had brought humankind to the threshold of a new age in which they would shape a better life for themselves, free from the ignorance of the past. In this process the scientific revolution would play an essential part. Scientists' had begun to penetrate nature's secrets, and this would inevitably continue until mankind was the absolute master of the environment. Optimism about the future and a belief in the limitlessness of progress sprang from this faith in reason and science. The *Encyclopédie* opened its article on political authority with the words:

"No man has received from nature the right to rule others. Liberty is a gift from heaven, and every individual of the race has the right to enjoy it as soon as he attains to the enjoyment of his reason."

Opposite: The tree of knowledge from the *Encyclopédie*. This is part of a large, complex diagram in which all the branches of human knowledge are related to each other. For example, mathematics leads on to geometry, mechanics, optics, statistics, and so on; physics leads on to astronomy, earth science, life science, and so on. The authors believed they were on the edge of solving all scientific problems by pure reason.

Below: A page from the *Encyclopédie* showing medical instruments and wound dressings.

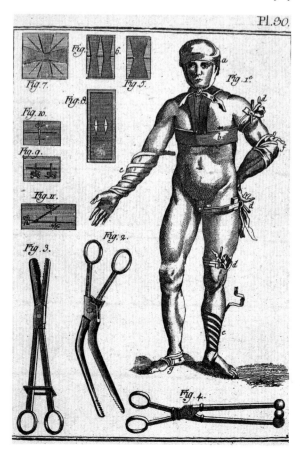

Pl.30.

Nature is appealed to as the highest authority in all things, and nature is understood through science. It is the scientist—the naturalist, the geographer, the anatomist—who can say what humans really are when stripped of their social trappings, and scientific truth therefore should underpin all social principles.

The *Encyclopédie* was the brainchild of two editors, Denis Diderot, a philosophical writer, and Jean Le Rond D'Alembert, a mathematician. Many of the articles were written by Diderot and D'Alembert, while others were contributed by great thinkers of the time, such as Montesquieu, Jean-Jacques Rousseau, Voltaire, d'Holbach, and the Marquis de Condorcet. Together these writers became known as the *Encyclopédistes.* Their critical articles on political topics, on religion, on history, and on philosophy provoked a storm of protest in prerevolutionary France, and the editors had to struggle against censorship and threats of arrest. A philosophical article on certainty based its arguments entirely on the scientific principle that knowledge can only be derived

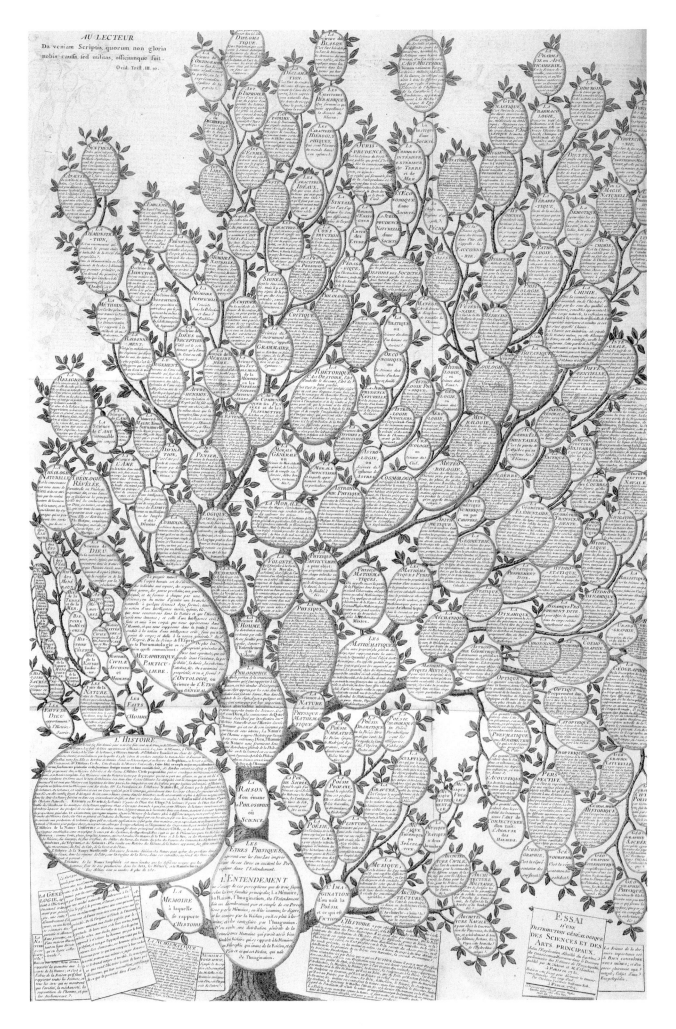

from experience and evidence: The revelations of the Bible were not mentioned. This kind of teaching was seen as subversive of religion.

RECORD OF SCIENCE

The *Encyclopédie* was democratizing knowledge on a massive scale. It described contemporary ideas on physics and chemistry, medicine, astronomy and earth science, and natural history. But in addition, it presented in words and in a series of illustrated plates all the crafts and technologies of the period—glassmaking, shipbuilding, the preparation of chemicals, clock design, weapon casting, the principles of the telescope, methods of surgery, the science of navigation—all these things and many more are explained in detail, making the *Encyclopédie* a complete record of the science of its time.

The writers of the *Encyclopédie* are seen by historians as preparing the way socially and intellectually for the French Revolution. Some of them, such as Condorcet, lived to welcome the Revolution and to predict that science would soon eliminate poverty, ignorance, and disease. In a word, he believed in the *perfectibility* of human life under the guidance of reason and science. Thinkers of the nineteenth century quickly learned that the gifts of science are more complex and much harder to control than that.

Below, Left and Right: Two more plates from the *Encyclopédie* showing spinning wheels (left) from Volume III and horn working from Volume IX.

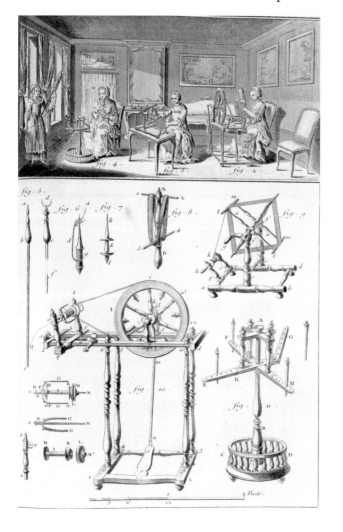

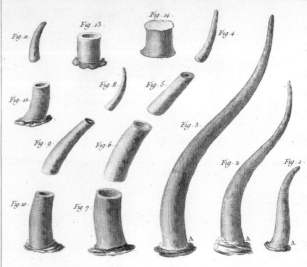

The Rationalization of Measures

All civilized societies, from the ancient Egyptians and Babylonians onward, had devised systems of measuring distance, weight, and volume. Any measurement system needs a set of units, and it needs standards that are the physical embodiment of those units from which weights or rulers can be copied. It is not surprising that the many different units devised in a multitude of societies show striking similarities: Any group of people will need small units, medium units, and large units; and the human body offers a basic sense of scale—a hand's breadth, an arm's length, the height of a person, and so on.

For as long as there has been commercial trading, there have been the means to weigh goods accurately. These are Babylonian weights and measures from the eighth century B.C. As well as the Babylonian bronze weights—the ram and the duck—there is also an Egyptian weight in the form of a bull's head from Amarna dating to the Eighteenth Dynasty, 1570–1293 B.C.

In the countries of medieval Europe a bewildering variety of customary weights and measures grew up, and even within one country regional variations in length and weight were well established. Many basic commodities, such as wheat or wool, were measured not by exact weights but by the volumes of traditional containers, such as the bushel, which could vary by several percentage points from place to place. It would require an enormous act of political will to rationalize any national system, and the opportunity finally arose in France in the late eighteenth century.

OBJECTIVE STANDARDS

The revolutionary government was anxious to rid France of all traces of the corrupt past, including the confused system of measurement. The other objection to the old systems was that they were arbitrary and irrational—they did not relate to any objective standard in nature, and this the rationalists of the eighteenth century found highly unsatisfactory. The rationalization of measures would become a small but important symbol of the way science could reform society, given the political will.

In fact, the age of science, a full century before the French Revolution, had already thrown up several new ideas about measurement. In 1670 Gabriel Mouton, a churchman from Lyons, had proposed that the basic unit of length should be directly related to the circumference of the Earth: It should be one minute of arc on a great circle. Given the Earth's circumference of approximately 25,000 miles, this would yield a unit of 1.15 miles. Mouton then proposed that this unit should be divided into seven subunits, each one-tenth the length of the previous one: a tenth, a hundredth, a thousandth, and so on. This was a highly logical decimal system that only lacked the political opportunity to put it into practice.

Late in 1790 the French National Assembly decided to begin the process of reforming the customary weights and measures along rational scientific lines. A high-level committee was appointed that included Laplace, the mathematician, Lavoisier, the chemist, and Condorcet, the philosopher. Letters were sent to England inviting cooperation in the scheme, but growing political hostility between the two countries prevented this. In the following year the committee recommended that the new standard measure should be one ten-millionth part of the meridian quadrant, that is, of the quadrant of the Earth from the north pole to the equator. The resulting unit was to be called the meter, and it was to be subdivided and multiplied according to the decimal system.

NEW RULE

The circumference of the Earth would thus be deemed to be 40 million meters—but exactly how long was one meter? In order to establish this figure, an arc of the meridian had to be precisely measured covering ten degrees from Dunkerque in western France to Barcelona in Spain. The deposed King Louis XVI authorized the work from his prison cell, and two mathematicians, Pierre Mechain and Jean-Baptiste Delambre, set out on a labor that would last seven years in the fog of war and revolution. Their results were calculated and embodied in a platinum standard of the meter, and the new system was finally enacted by the French government in 1799 "for all people for all time."

At the same time, new rational measures of weight and volume had been devised. The basic unit of weight, the gram, was equal to the weight of one cubic centimeter of pure water at a temperature of 4°C (39°F), when it is at its maximum

Ramsden's balance. Jesse Ramsden (1735–1800) was an English instrument maker who improved optical and survey instruments.

density; a definitive kilogram was manufactured from platinum. A liter was defined as the volume of a cube whose sides were all 10 centimeters, that is, 1,000 cubic centimeters. The new measures were considered the first in human history to be rational, to have a natural basis.

French political power carried the new decimal measures throughout Europe, but its acceptance by the common people was slow—indeed, in France itself in 1812 Napoleon permitted a partial return to the old units. Not until 1837 was the decimal system made the exclusive official system of France.

Neighboring Britain had refused to consider the French scheme for political reasons, but many people in America, including Thomas Jefferson, were in favor of it. The great problem was that Britain was still America's greatest trading partner, and commercial interests feared the confusion that would result if America went decimal. And so the opportunity passed. Much later, in the 1890s, Britain adopted metric measures alongside its traditional ones. All the international scientific community now uses metric measures, one of the most visible legacies of the Age of Reason; but many people still prefer their customary, irrational, age-old measures.

Accurate mapping requires accurate instruments. This 100-link, 100-foot-long brass and steel chain (made by Jesse Ramsden) was used in summer 1784 in the primary triangulation of Great Britain. Beneath it lies the plan showing the base measurement on Hounslow Heath, London. From this base General William Roy (1726–90) of the Royal Engineers was able to make accurate measurements of the country for the first time, and his work would lead on to the first Ordnance Survey maps of Britain.

Retrospective:
Science in the Age of Reason

The eighteenth century was the period when European science came of age, when a knowledge of the principal scientific concepts became part of its familiar language, and when the scientific idea of truth—as derived from experience and verifiable by experiment—was accepted as philosophically valid. The two essential sciences of the age were astronomy and natural history, and they made major advances in technique and in theory.

As the status of scientific thought rose higher, the question inevitably arose of its effect on older beliefs, especially religious ones. In England Newtonian physics was generally felt to support a religious view of the universe: The Earth and the heavens were seen as a vast, divinely created mechanism moving eternally in perfect order.

The supreme scientific image of the age was the "orrery," the clockwork model of the solar system that showed the planets' movements around the Sun. These ingenious toys were made for gentlemen's libraries, and they became a form of social entertainment. They flattered their audience's sense of having

Above: An early (around 1710) compound microscope made of brass, oak, and vellum by John Marshall of London.

Right: An orrery of around 1740 made by Thomas Wright for "ladies and gentlemen rather than noblemen or princes." It belonged to a doctor named Stephen Demainbray (1710–82), a science lecturer. The orrery shows the Sun, Earth, Moon, Mercury, and Venus. The Earth can be turned on its axis directly, while the handle is used for showing annual rotation. The outer ring shows the calendar, the signs of the zodiac, and the points of the compass.

penetrated one of the deepest of nature's workings; and where astronomy had gone, surely other sciences would follow in time, revealing the laws of biology, medicine, earth science, chemistry, and physics. But the sense of *design* everywhere in nature left these scientists in no doubt that the universe was sustained by God.

In France this optimism about the future of science was, if anything, greater still, but it tended to inspire a materialism which grew into outright atheism.

CONTRASTING VIEWS

Among the writers of the *Encyclopédie* Baron d'Holbach was the most outspoken. To him nature had to be seen as a self-sustaining mechanism that needed no divine intelligence to control it. "Theology," wrote d'Holbach, "is but the ignorance of natural causes reduced to a system." The gods, whether of ancient times or modern, were merely labels for people's ignorance of nature's laws. As science progressed, it would inevitably demythologize the universe. This view of science gave it the power to criticize and overturn traditional forms of thought and belief. These two contrasting views of science, the English and the French, both showed the effect of science on traditional belief systems.

In the life sciences the great problematic issue was the fixity of

It may look like an alchemist's lair but it is in fact a model of the laboratory interior in which Frederik Cronstedt discovered nickel in 1751. It's typical of the type of laboratory used by metallurgical chemists of the time.

Above: The workshop of James Watt (see above, pages 42–45) as it was in 1790.

Opposite, Above Left: Benjamin Franklin, American printer, philosopher, and scientist. This is one of a series of tiles produced by Doulton and Co. that were exhibited in Edinburgh, Scotland, in 1886.

Opposite, Above Right: Volta's letter to Sir Joseph Banks of the Royal Society of London in 1800 detailing the first practical battery. It produced just over one volt for every set of disks.

Opposite, Below: The octagon room at Greenwich Observatory.

species. Again, it was the French naturalists of the school of Buffon who went furthest in rejecting the simplistic biblical doctrine that all species had been created at one moment, replacing it with the more dynamic view that nature has experienced a long process of change.

The same conclusion was reached by scientists who studied the fabric of the Earth itself: Its mountains, rivers, rocks, and lakes had surely been formed over countless thousands of years. Exactly how these changes operated Buffon and others could only speculate, but the time scale of the Earth's existence must, they felt, be extended well beyond the few thousand years that the Bible suggests. Thus science was offering people a radically new perspective on their world.

The most intractable problems still lay in the science of matter, in physics and chemistry. What held matter together, and how did it dissolve and change its nature? Did it contain mysterious components such as phlogiston or electricity; and if so, how could they be isolated? Less glamorous than the mysteries of life or cosmic space, the science of matter had largely resisted all the assaults of the Age of Reason. But by the end of the century signs of a new beginning were evident. The establishment of classical physics and chemistry awaited a new generation before it in turn would become the central science of the age.

VOL. I. — N. XXII.

BRANO DI UNA MINUTA DELLA LETTERA A SIR J. BANKS (da Cart. Volt. J 68).

PROSPECTUS INTRA CAMERAM STELLATAM.

Glossary

Captions to illustrations on page 72.

Above: James Hutton—see pages 46–48.

Below: Title page of the second volume of Benjamin Franklin's complete works. A polymath of considerable talent, Franklin was an inventor, scientist, printer, publisher, and statesman.

Observing the Moon around 1700. This hand-colored aquatint by Sergent is of French author Bernard le Bovier de Fontenelle (1657–1757). Fontenelle was secretary to the French Academie des Sciences from 1697, and later president.

anomalies (pl of anomaly) irregularities; something different or not easily classified.

anther the part of the stamen that produces pollen.

atheism the teaching that there is no deity.

binomial system in which two Latin words give a unique label to any plant.

centrifugal force the force that impels things out from a rotating center.

class a category in biological classification.

condenser an apparatus in which gas or vapor is condensed.

conductivity the quality of conducting or transmitting electricity.

corolla the part of a flower consisting of separate or fused petals.

coulomb the standard unit of charge, one amp per second.

ecology the pattern of relationships between organisms and their environment.

embryology branch of biology dealing with embryos and their development.

genus a biological classification ranking between the family and the species.

germination beginning to sprout or develop.

hermetically sealed airtight.

horsepower a unit of power equal to 746 watts.

life science branch of science dealing with living organisms.

lymph liquid resembling blood plasma and containing white blood cells but normally no red blood cells.

meridian an imaginary circle on the face of the Earth

passing through the
poles.

mechanistic theories
theories that the
universe is a
machine.

order a biological
classification
ranking above the
family and below
the class.

orrery working models
of the solar system,
with planets that
revolved around a
sun.

orthodoxy the
established belief.

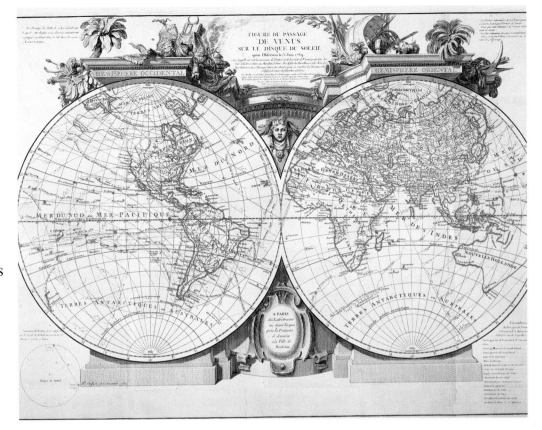

parallax the angular difference in direction of a celestial body
measured from two points on the Earth's orbit.

pathology the deviations from the normal that characterize disease

phlogiston Georg Stahl's theory about combustion.

photosynthesis formation of carbohydrates from carbon dioxide
and hydrogen in water in chlorophyll-containing tissues of plants
when exposed to light.

physiology the organic processes of an organism; a branch of
biology studying living matter.

rationalist a person who relies on reason as a basis for finding
religious or scientific truth.

resistance the opposition offered by a substance to the passage of
an electric current.

species a biological classification below the genus comprising
related organisms capable of interbreeding.

stamens the pollen-producing male organ of a flower.

taxonomy the classification of plants and animals according to their
relationships.

virus infectious agent that grows and multiplies only in living cells
and causes disease.

vortices (pl of vortex) a whirling mass of water or air, such as a
whirlpool or tornado.

**This map has superimposed on it
the path for the best observation
of the transit of Venus across the
sun on June 5, 1769.**

**Captions to illustrations on page
73.**

**Above Left: Carl Linnaeus, in
whose honor the Linnaean society
was founded in 1788, was the
deviser of modern nomenclature
for plants and animals. See pages
20–23.**

**Center Right: The eighteenth
century saw the first precision
machinery constructed—such as
this lathe. The foundations for the
Industrial Revolution were laid in
this century.**

**Bottom: A nineteenth-century
reprint of Thomas Savery's patent
for "Machinery for raising water,
giving motion to mills, etc." See
page 40.**

PERIOD	1700	1710	1720	1730	1740	1750

WORLD EVENTS

1701–14 War of Spanish Succession (Marlboroughs War).
1701–20 Great Northern War—Scandinavia, Russia, Poland, Saxony.
1703 Founding of city of St. Petersburg.
1707 Great Britain formed from union of England, Wales, and Scotland.
1715 Death of Louis XIV.
1740-48 War of Austrian Succession.

1740 Rise of Prussian power under Frederick the Great.

1750 Industrial Revolution begins in England with ironworking and steam power.

SCIENCE

1700 Halley charts world magnetic variation.
1700 Stahl promotes animism and theory of phlogiston.
1704 Newton's *Opticks* published.
1705 Halley's analysis of comet orbits.
1712 Newcomen's steam engine.
1718 Halley discovers motions of the stars.
1727 Death of Newton.
1730 Linnaeus's system of plant classification.
1743-57 Haller's *Icones Anatomicae* ("Images of Anatomy").
1745 Maupertuis of Canefanino reports on the shape of the Earth.
1746 Mussenbroek invents electrical condenser.
1746–47 Franklin shows lightning is a form of electricity.
1747 Albinus's *Atlas of Anatomy*.
1748 La Mettries's *L'Homme Machine* ("The Human Machine") published.
1749 Buffon's *Histoire Naturelle* ("Natural History") begins publication (completed 1789).

ART & CULTURAL EVENTS

1700–30 Augustan literature in England: Dryden, Swift, Pope
1700–50 Baroque music: Bach & Handel
1710 onward development of English novel: Defoe, Fielding, etc.
1730–50 Hogarth's satirical paintings
1730–50 Canaletto's Venetian paintings
1750 Rococo art in Europe: Tiepolo's palace at Würzburg; Sans-Souci Palace, Potsdam

1750　　1760　　1770　　1780　　1790　　1800

1755 Great Lisbon earthquake.
1756–63 Seven Years War: Britain destroys French power in North America and India.
1768–79 Captain Cook explores the Pacific Ocean.
1776–83 American Revolutionary War.
1788 British colony of Australia founded.
1789 George Washington elected first president of United States.
1789 French Revolution.
1798 Rise of Napoleon to power.

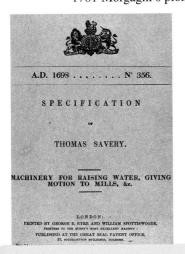

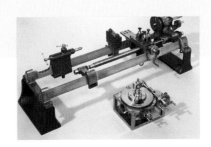

1750 La Caille charts southern skies.
1751 *L'Encyclopédie* starts publication.
1760 James Watt redesigns the steam engine.
1761 Morgagni's pioneer work on pathology.
1781 Ingenhousz discovers photosynthesis.
1784 Messier's catalogue of nebulas.
1785 Coulomb measures electrical force.
1790 Galvani's electrical experiments.
1795 Hutton's *Theory of the Earth* published.
1795–1815 Laplace working on *Celestial Mechanics* and
　　　　on the nebular theory.
1797 Jenner develops smallpox vaccine.
1799 Decimal system finalized in France.
1800 Lamarck begins work on evolution theory.

A.D. 1698 . . . . Nº 356.

SPECIFICATION
of
THOMAS SAVERY.

MACHINERY FOR RAISING WATER, GIVING
MOTION TO MILLS, &c.

LONDON:
PRINTED BY GEORGE E. EYRE AND WILLIAM SPOTTISWOODE,
PRINTERS TO THE QUEEN'S MOST EXCELLENT MAJESTY:
PUBLISHED AT THE GREAT SEAL PATENT OFFICE,
25, SOUTHAMPTON BUILDINGS, HOLBORN.

1750 onward The European Enlightenment: Voltaire's satires; Rousseau's philosophy of nature; Hume's natural philosophy;
　　　　Gibbon's critical history.
1751 Gray's "Elegy in a Country Churchyard."
1755 Johnson's *Dictionary*.
1760–1800 Neoclassical art and architecture in England: Adam houses; Reynold's paintings.
1770–90 Music of Hayden and Mozart.
1791 Paine's *Rights of Man* published.
1792 Wollstonecroft's *Rights of Women* published.

Resources

FURTHER READING

There is a wealth of books published on the history of science, particularly biographies of great scientists. The following list includes many large works that contain many further resources.

Adams, F.D.: *The Birth and Development of the Geological Sciences*; Dover Publications, 1955.

Bowler, P.: *Evolution: The History of an Idea*; University of California Press, 1998.

Bowler, P.: *The Norton History of Environmental Sciences*; W.W. Norton & Co., 1993.

Boyer, C. & Merzbach, U.: *A History of Mathematics*; John Wiley & Sons Inc., 1989

Brock, W.H.: *The Fontana History of Chemistry*; Fontana Press, 1992.

Butterfield, H.: *The Origins of Modern Science*; Free Press, 1997.

Clagett, M.: *Greek Science in Antiquity*; Dover Publications, 2002.

Cohen, I.B.: *Album of Science: From Leonardo to Lavoisier*; Charles Scribner's Sons, 1980.

Cohen, I.B.: *The Birth of a New Physics*; W.W. Norton, 1985.

Crombie, A.C.: *Augustine to Galileo: The History of Science AD400–1650*; Dover Publications, 1996.

Crombie, A.C.: *Science, Art and Nature in Medieval and Modern Thought*; Hambledon, 1996.

Crosland, M.: *Historical Studies in the Language of Chemistry*; Heinemann Educational, 1962.

Eves, H.: *An Introduction to the History of Mathematics*; Thomson Learning, 1990.

Gillispie, C.C. (ed.): *Concise Dictionary of Scientific Biography*; Charles Scribner's Sons, 2000.

Gillispie, C.C.: *Genesis and Geology*; Harvard University Press, 1996.

Hallam, A.: *Great Geological Controversies*; Oxford University Press, 1983.

Ihde, A.J.: *The Development of Modern Chemistry*; Dover Publications, 1983.

Jaffé, B.: *Crucibles: The Story of Chemistry from Alchemy to Nuclear Fission*; Dover Publications, 1977.

Jungnickel, C. & McCormmach, R.: *Intellectual Mastery of Nature: Theoretical Physics from Ohm to Einstein*; University of Chicago Press, 1986.

Koyré, A.: *From the Closed World to the Infinite Universe*; The Johns Hopkins University Press, 1994.

Kuhn, T.: *The Copernican Revolution: Planetary Astronomy in the Development of Western Thought*; Harvard University Press, 1957.

Lindberg, D.C.: *The Beginnings of Western Science*; University of Chicago Press, 1992.

Porter, R. (Ed.): *The Cambridge Illustrated History of Medicine*; Cambridge University Press, 1996.

McKenzie, A.E.E.: *The Major Achievements of Science*; Iowa State Press, 1988.

Morton, A.G.: *A History of Botanical Science*; Academic Press, 1981.

Nasr, S.H.: *Islamic Science—An Illustrated Study*; London, 1976.

North, J.D.: *The Fontana History of Astronomy and Cosmology*; Fontana Press, 1992.

Olby, R. (et al.): *A Companion to the History of Modern Science*; Routledge, 1996.

Parry, M. (ed.): *Chambers Biographical Dictionary*; Chambers Harrap, 1997.

Porter, R.: *The Greatest Benefit to Mankind: a Medicinal History of Humanity from Antiquity to the Present*; HarperCollins, 1997.

Roberts, G.: *The Mirror of Alchemy*; British Library Publishing, 1995.

Ronan, C.A.: *The Cambridge Illustrated History of the World's Science*; Cambridge University Press, 1983.

Ronan, C.A.: *The Shorter Science and Civilisation in China*; Cambridge University Press, 1980.

Selin, H. (ed.): *Encyclopedia of the History of Science, Technology and Medicine in Non-Western Cultures*; Kluwer Academic Publishers, 1997.

Uglow, J.: *The Lunar Men*; Faber and Faber, 2002.

Van Helden, A.: *Measuring the Universe: Cosmic Dimensions from Aristarchus to Halley*; University of Chicago Press, 1985.

Walker, C.B.F. (ed.): *Astronomy before the Telescope*; British Museum Publications., 1997.

Whitfield, P.: *Landmarks in Western Science: From Prehistory to the Atomic Age*; The British Library, London, 1999.

Whitney, C.: *The Discovery of Our Galaxy*; Iowa State University Press, 1988.

THE INTERNET

Websites relating to the history of science break down into four types:

- Museum sites that offer some history and artifact photography. This is often the easiest way to visit international sites or those of states too far away to get to in person.
- College or other educational establishment sites that often provide online learning or study resources.
- General educational sites set up by enthusiasts (often teachers) and historians.
- Societies or clubs.

Examples of these three types of website include:

Museums
http://www.mhs.ox.ac.uk/
Museum of the History of Science, Oxford, England.
Housed in the world's oldest surviving purpose-built museum building, the Old Ashmolean.

http://www.mos.org/
Museum of Science, Boston.

http://www.msichicago.org/
Museum of Science and Industry, Chicago.

http://www.lanl.gov/museum
Bradbury Science Museum, a component of Los Alamos National Laboratory.

http://www.si.edu/history_and_culture/history_of_science_and_technology/
Smithsonian Institution site.

http://www.sciencemuseum.org.uk/
National Museum of Science and Industry, London, England.

http://galileo.imss.firenze.it/
Institute and Museum of the History of Science, Florence, Italy.

http://www.jsf.or.jp/index_e.html
Science Museum, Tokyo.

Colleges or institutions
http://sln.fi.edu/tfi/welcome.html
Franklin Institute with online learning resources and study units.

http://www.fas.harvard.edu/~hsdept/
Department of the History of Science of Harvard University.

http://www.hopkinsmedicine.org/graduateprograms/history_of_science/
Department of the History of Science, Medicine and Technology at Hopkins.

http://www.lib.lsu.edu/sci/chem/internet/history.html
Louisiana State University provides excellent history of science internet resources and links.

http://dibinst.mit.edu/
The Dibner Institute is an international center for advanced research in the history of science and technology and located on the campus of MIT.

http://www.mpiwg-berlin.mpg.de/ENGLHOME.HTM
Max Planck Institute for the History of Science

http://www.princeton.edu/~hos/
History of Science @ Princeton.

http://www.astro.uni-bonn.de/~pbrosche/hist_sci/hs_sciences.html
History of sciences from Bonn University, Germany, including indexes on the history of astronomy, chemistry, computing, geosciences, mathematics, physics, technology.

Educational sites
http://echo.gmu.edu/center/
ECHO—Exploring and Collecting History Online—provides a centralized guide for those looking for websites on the history of science and technology.

http://www.wsulibs.wsu.edu/hist-of-science/bib.html
Provides reference sources in the form of bibliographies and indexes.

http://dmoz.org/Society/History/By_Topic/Science/Engineering_and_Technology/
Open Directory Project providing bibliography and links.

http://orb.rhodes.edu/
ORB—the Online Reference Book—provides textbook sources for medieval studies on the web. It includes the Medieval Technology Pages—providing information on technological innovation and related subjects in western Europe—and Medieval Science Pages, a comprehensive page of links to medieval science and technology websites.

http://www.fordham.edu/halsall/science/sciencesbook.html
This page provides access to three major online resources, the Internet Ancient History, Medieval, and Modern History Sourcebooks.

http://www2.lib.udel.edu/subj/hsci/internet.htm
The University of Delaware Library provides an excellent guide to Internet resources.

Societies
www.hssonline.org
History of Science Society provides for its members the History of Science, Technology, and Medicine Database—an international bibliography for the history of science, technology, and medicine.

http://www.chstm.man.ac.uk/bshs/
British Society for the History of Science.

Set Index